EARTHDANCE

EARTHDANCE

LIVING SYSTEMS IN EVOLUTION

Elisabet Sahtouris

iUniversity Press
San Jose New York Lincoln Shanghai

EarthDance
Living Systems in Evolution

iUniversity Press
an imprint of iUniverse.com, Inc.

For information address:
iUniverse.com, Inc.
5220 S 16th, Ste. 200
Lincoln, NE 68512
www.iuniverse.com

ISBN: 0-595-13067-4

Printed in the United States of America

To my planet and its people

Dancing is surely the most basic and relevant of all forms of expression. Nothing else can so effectively give outward form to an inner experience. Poetry and music exist in time. Painting and architecture are a part of space. But only the dance lives at once in both space and time. In it the creator and the thing created, the artist and the expression, are one. Each participant is completely in the other. There could be no better metaphor for an understanding of the...cosmos.

We begin to realize that our universe is in a sense brought into being by the participation of those involved in it. It is a dance, for participation is its organizing principle. This is the important new concept of quantum mechanics. It takes the place in our understanding of the old notion of observation, of watching without getting involved. Quantum theory says it can't be done. That spectators can sit in their rigid row as long as they like, but there will never be a performance unless at least one of them takes part. And conversely, that it needs only one participant, because that one is the essence of all people and the quintessence of the cosmos.

—Lyall Watson, *Gifts of Unknown Things*

Thank you...

Special thanks to Jim Lovelock and Lynn Margulis for the original inspiration to write this book and for their encouragement as it came into being, also to Teddy Goldsmith for creating the Gaia Seminars in Cornwall. My deep appreciation to Dave Ratcliffe and Rebecca Lord for putting the book on the Web while it was out of print, and to Bruce Bigenho for his generous efforts with the second edition. My gratitude extends as well to Nancy Larson for the original cover photo and to my son, Philip LaVere, for the cover design.

Contents

Foreword

The Gaia hypothesis, now accorded the status of Gaia theory, is maturing with experience and the tests of time, not unlike the humans of this book. It is spurring a great deal of scientific research into the geophysiology of our living planet. It is also spurring philosophic conceptions of what it means to our species to be part of a living planet. Some of these conceptions stay carefully within the accepted limits of science; others have a religious bent. Most, especially environmentalist conceptions, advocate for humanity, being primarily concerned with human survival. A few, taking a clue from my partner Lynn Margulis and myself, advocate for the planet and the much maligned microbes with which the Gaian system originated and which continue to do its basic work.

Elisabet Sahtouris' conception integrates scientific Gaian evolution with the human search to connect with our roots, inspiring us to learn from billions of years of Gaian experience in the self-organization of workable living systems. It is well balanced between advocacy for the planet and advocacy for humans, placing the onus on humans to recognize the lack of maturity involved in believing we can manage the planet, and to learn instead to follow its lead in organizing ourselves.

Elisabet gives us valuable insights as she draws parallels between the evolution of cells and the evolution of human society, pointing out the contrast between the healthy organization of cells, bodies, and biosystems on the one hand and the unhealthy organization of economics and politics in human society on the other. While she argues that our social evolution is not as much under our control as we like to think, she warns us that our survival depends on our meeting the evolutionary demand to transform competitive exploitation into cooperative synergy.

On the whole, her advice makes sense because she herself has taken the trouble to learn directly from nature as well as from the growing store of scientific knowledge about nature. I began the preface to my own book *The Ages of Gaia* by saying that the place in which it was written was relevant to its understanding. Living and working in the Devonshire countryside, far from universities and large research organizations, makes me an eccentric as a scientist, but, as I said, it is the only way to work on an unconventional topic such as Gaia. When I met Elisabet, having accepted her invitation to trace Gaia's roots in Greece, I recognized her as a kindred spirit. She had abandoned academia for a simple lifestyle in the kind of natural setting that brings one closer to understanding what our planet and our species are all about; she was free to develop her own conception of Gaia through a synthesis of scientific knowledge and personal experience of nature. To my surprise, she expressed some concern, some guilt, at having abandoned her profession of science for a pleasant existence in a forest overlooking the sea, the kind of forest that had been home to her in childhood, where she could work out the meaning of things for herself. As I read her work in progress, I was able to assure her she could never have done anything comparable in a constrained academic setting.

In the intervening years, even in the short time since I wrote my own words about Gaia being an unconventional topic, less eccentric scientists than I have declared Gaia more conventional, meaning that Gaia theory is now recognized as a legitimate and fruitful basis for scientific investigation and is thus being brought into the scientific fold. In our first account of Gaia as a system neither Lynn Margulis nor I fully understood what it was we were describing. Our language tended to be anthropomorphic and, especially in my first book, Gaia, poetic. Not surprisingly, some scientists misunderstood our intentions, but over time we developed a clearer version, which became Gaia theory. This theory sees the evolution of the material environment and the evolution of organisms as tightly coupled into a single and indivisible process or domain. Gaia, with its capacity for homeostasis, is an emergent property of this domain.

As the title of one article in *Science* put it, "No Longer Willful, Gaia Becomes Respectable." This means that Gaia scientists are constrained by bureaucratic forces, by the pressures of tenure, and by the tribal divisions and rules of scientific disciplines. That, in turn, means we need some antidote to the inevitable separations and constraints. We need independent synthesizers and visionaries who can make sense of the data produced by the scientific establishment and present it to us in ways that make our living planet real to us within the Gaian context and thus give meaning to our own lives and those of our children and grandchildren.

This is what Elisabet Sahtouris' work means to me, for she comfortably integrates the traditionally separated domains of biology, geology, and atmospheric science to show us the evolution of our living planet and our own roots within it. She then inspires us on ethical grounds to learn from this planetary organism of which we are part, showing us how we can mature as a species well integrated into the larger dance of life.

Elisabet uses the metaphor of dance effectively for its concepts of improvisation and evolution, the creation of order from chaos, the myriad patterns that can be created from a few basic steps. I am myself an inventor of scientific instruments, and so it is second nature to me to think in terms of mechanical and mathematical models. Cybernetic models have proved especially useful in my work of demonstrating how Gaian homeostasis, such as maintaining the Earth's temperature, might work. Yet I quite agree with Elisabet that any model we make of nature is at heart metaphorical in that it begins with some image or formula familiar to us humans and used to represent the complexities of nature in simple, understandable, and useful ways. No metaphor should be mistaken for reality, and perhaps a variety of metaphors is insurance against the temptation to do so. I am increasingly impressed by scientists and philosophers who find non-mechanical metaphors for natural systems useful in interpreting Gaia theory.

Elisabet's analysis of science reflects a trend that may well make science in the near future as unrecognizable as today's science would be to the ancients. She does well to remind us that science is a human activity that evolves, a living system in which conservatism should be balanced by healthy controversy. After all, as she so well describes, all Gaian systems are forever busy working out their cooperation through conflicting interests, their unities through diversity.

The optimistic view this book radiates, that despite our errors and immaturities we can still become a healthy species within a healthy planet, is much needed in this age of doomsday predictions. Though time is growing short in our continued destruction of forests, atmospheres, and other critical Gaian systems, nothing would make me happier personally than to see Gaia theory useful in bringing about a better world for Gaia and her people.

—*James E. Lovelock*

A Note from the Author

This book is a work of philosophy in the original sense of a search for wisdom, for practical guidance in human affairs through understanding the natural order of the cosmos to which we belong. It bears little resemblance to what we have come to call philosophy since that effort was separated from natural science and became more an intellectual exercise in understanding than a practical guide for living.

To find meaning and guidance in nature, I integrated my personal experience of it with those scientific accounts that seemed to best fit it. From this synthesis, meaning and lessons for humanity emerged freely. I wrote the original version in the peaceful, natural setting of a tiny old village on a small pine-forested Greek island, where I could consider the research and debates of scientists, historians, and philosophers, then test them against the natural world I was trying to understand.

Putting into simple words the specialized technical language of scientists and winding my way through labyrinths of philosophic prose, I gradually simplified the story of the origins and nature of our planet within the larger cosmos, and of our human origins, nature, and history within the larger being of this planet.

The Gaia hypothesis, now Gaia theory, of James Lovelock and Lynn Margulis—the theory that our planet and its creatures constitute a single self-regulating system that is in fact a great living being—is the conception of physical reality in which my philosophy is rooted. Quite simply, it makes more sense on all levels—intuitive, experiential, scientific, philosophical, spiritual and even aesthetic and ethical—than any other conception I know. And I have come to believe, in the course of this work, that this conception contains profound and pressing implications for all humanity.

To ensure that my vision of evolution and history would stay simple and in clear focus, I kept telling its essence and more than a few of its particulars in something of the style of an ancient storyteller during many social evenings among my Greek village friends. I also wrote the story for children before I set about an adult version. To my surprise, these deliberate exercises in simplicity proved more difficult than writing for professional audiences, for in stripping our intellectual language to the essence of what is being said, we must be very sure that essence is really there, really coherent. Science has been a process of differentiating our knowledge into an incredible wealth of precise details, but these details become ever more disconnected from one another and cry out for integration into coherent wholes. I have no doubt I will be accused of oversimplification, and perhaps rightly so, as one pays for scope in lack of detail and precision.

Friends and colleagues have asked me now and then why I insist on dealing with all evolution, even all the cosmos, to discuss human matters; why I don't narrow my scope to workable proportions. My answer is that context is what gives meaning, and a serious search of context is an ever-expanding process leading inevitably to the grandest context of all: the whole cosmos. As the nested contexts for the human story—especially the context of evolution—became clearer to me, they revealed a simple but elegant biological vision of just why our human condition has become so critical and what we might do to improve it.

Other people ask why I'm so eager to save humanity when it is proving such a social and ecological disaster. To this I can only answer that, as far as I can see, every healthy living being or system in nature has evolved survival oriented behavior, and I do not exclude myself from this natural health scheme. Of course my purpose is to show how we are straying from this course, so that we may correct the deviations.

I can no more proclaim the worldview arising from my work "reality" than can any particular philosopher working at creating a meaningful worldview in any particular place and time, drawing on the scientific and historical knowledge of that place and time. Philosophy is an intensely personal search that one hopes will have relevance to others, will be validated by their experience, will offer them some insight and guidance, or will at least stimulate them in their disagreement to search further on their own.

Yet a work of philosophy also reflects the broader context and search of a culture at a particular stage, and the biological evolutionary viewpoint of this book reflects a broadly emerging pattern of search for our origins and direction in nature—a reawakening of that search begun by the original pre-Socratic philosophers, indeed that goes further back to the roots of religion—the search for *religio*, for "reconnection" with our origins in the nature or cosmos that gave rise to us and within which we continue our co-creation.

Paradoxically, our self-imposed separation from nature by way of an 'objective' mechanical worldview during the past few millennia has led to the scientific knowledge that makes it possible to understand and reintegrate ourselves into nature's self-organization patterns. It has also brought us to a stage of technology that permits us to share our discoveries and our understanding planet-wide in no time at all, to work together as a body of humanity with hope of transcending our present crisis in a far healthier and happier future for ourselves and all the rest of Earthlife.

Although the original version of this book was done in relative isolation and without funding, I am indebted and profoundly grateful to many teachers and friends, from the forest creatures with whom I spent my earliest years to Jim Lovelock and Lynn Margulis, who have not only informed and inspired me in this work, but who gave me invaluable encouragement, confidence, and opportunities in seeing the work through.

As this edition goes to press, scientists have recognized that we are well into the sixth great extinction of species—the first caused by a single species, and proceeding more rapidly even than the last one, which eliminated the great dinosaurs sixty million years ago because Earth's climate changed dramatically under the impact of a huge meteor in the Caribbean basin.

There is no doubt that we humans continue creating the chaos of ongoing disaster and denial. As I say in Chapter 19, Onondaga Chief Oren Lyons, at the Earth Summit known as Rio '92, reminded us that the passengers of the Titanic refused to believe that marvel of modern technology could go down on its maiden voyage. It did, of course, go down, as its extremely popular and timely Hollywood version reminded us. We may be a true biological marvel as a hi-tech human species, but we have truly gotten ourselves into serious trouble.

A healthy world for all cannot easily rise from total destruction; rather it must be formed now, in the midst of the chaos we create. Such a "new world order," I am again and again reminded by the indigenous elders I have listened to intently for their deep understanding of sustainability, must be based on a very old world order—on the laws of nature as indigenous people understand them, on laws they have been trying to teach us for a very long time: laws of balance, harmony, of giving back in full measure for all you take; laws designed to insure survival at least seven generations into the future.

The conclusion reached in this book, that we humans as a species must learn quickly to fit our lifestyles harmoniously into the rest of nature, is what led me to seek out indigenous knowledge between editions. Indigenous peoples never saw themselves as anything *but* an integral part of nature, and so they tend to know much more about that than do industrial peoples. Once, I listened to Jeannette Armstrong, a wise woman of the Okinakan nation, which still lives traditionally, speaking in detail about her peoples' understanding of nature. It was precisely the understanding I had gained in the course of writing this book far off on a Greek island— confirmation to me that I had gotten it right, for her people had the credibility of thousands of years of careful and scientific observation.

The immense knowledge of nature, the coherent philosophies and the non-technological achievements of indigenous people impressed me deeply. They have observed us far more carefully than we have them. Their conscious choice not to develop technological consumer societies gave me a more balanced view of human life and some valuable insights I have shared in several new chapters. One of these insights— that there can no more be one true science than one true religion—was difficult to share with fellow scientists of my industrial culture. Almost invariably, they responded, "You mean indigenous *knowledge*; they don't have *science*, there is only *one* science." I have therefore taken some care to show that indigenous people do indeed have science, by our own definitions, as a deep aspect of their cultures (see Chapter 19).

The great effort of industrial culture to fragment our world, to separate science, religion, art, economics, politics and other social practices, has long seemed to me very costly in blinding us to their interrelations. Today this is expressed in such problems as the difficulty of integrating the *economy* with *ecology*, two words meaning, in their original Greek, the organizational design and the operating principles of a household. Clearly they should never have been separated! How could it have happened? As Janine Benyus pointed out in a speech at a Bioneers conference, we assigned one group of people—biologists—to study how other species make a living, and another unrelated group of people—economists—to determine how humans make a living. Only now do we see interest in living systems enter the world of business.

Indigenous people have also taught me that good science can be done without tearing it out of the seamless and sacred fabric of life. They have always known this is a participatory universe, which Western scientists only now acknowledge. We simply cannot observe it without changing it. Indigenous people understand science and spirituality as aspects of the same reality—an intelligent, conscious continuum with physical and non-physical aspects. They are aware that all parts and aspects of nature are in constant non-physical communication. In Western science, physicists only now discover the deep connectedness and dialogue of everything through concepts of non-locality and zero-point energy.

One crisp cool day in a cornfield on the barren Hopi reservation in Arizona, I watched Martin Gashweseoma—now almost the only traditional Hopi elder still alive—kneeling in the dry earth beneath a brilliant blue sky, picking dried ears of blue corn from the stubby plants rustling in a cold late fall wind. Martin continues to live in the sacred way, with only the digging stick given by the Great Spirit, Maasau, along with instructions for living in peace and simplicity. He stood up to greet me and began speaking of the eviction of the faithful Hopi from Old Oraibi in 1906 with only what they could carry, of his uncle Yukiuma who led his people like Gandhi on this exodus, even going to the White House to plead their cause, of the sacred stone tablets his uncle later entrusted to him, of the way they were taken away, of the Day of Purification the white man, Bahanna, is bringing on, with all its suffering as the world becomes desert...

What he said was familiar, as I had been working with the Hopi and other Indians for years by this time, but it took on new significance as it burned into my heart on that crisp, clear fall day, the azure sky blazing behind him as we talked. Three men who had brought me to the field stood behind me and never interrupted; Martin did not take his eyes from mine during our long interchange. It was an experience of total undivided attention I, as a woman, had never experienced from men. The intense energy flowing between Martin and myself created a dense whirlpool tangible even to me, a person normally insensitive to such things. A whirlpool, as I say in this book, is a living entity, and Martin wove such an entity.

Anguish flowed through me at his despair. He spoke of his and other elders' failure to reach the White Brother—our dominant culture—with the Hopi Prophecy, and of how even the Hopi were abandoning their traditions, their cornfields. The Hopi prophecy, discussed at the beginning of Chapter 19, says the world as we know it will end if the White Brother does not heed the Sacred Way of the Red Brother and share his mission to develop technology in that spirit.

His truth—the need for cooperation between the ways of indigenous and industrial peoples to build a sustainable world—is vital to our survival. I found this same truth over and over again in many teachings I have gained from indigenous peoples in many places. I explored this truth in many contexts, from presidential commission dialogues on a sustainable human future in Washington DC to traditional villages in the Peruvian Andes, where I spent a whole year studying the cosmology and science of ancient Andean cultures, and now in the corporate world of multinationals, the most powerful organizations humanity has yet devised.

This corporate world, which, along with science and technology, is often blamed for current crises, is suddenly in crisis itself because of a dramatic new development on the human scene: the Internet. From my perspective as an evolution biologist, this WorldWideWeb of information exchange is a kind of fractal biology repeat pattern of the first version, built by bacteria billions of years ago, as we see in Chapter 4. And just like its ancient counterpart—still in existence among bacteria worldwide today—it is a self-organizing living system. This Internet is also an increasingly massive exercise in inventing democracy anew, and quite possibly the most significant influence in changing human society at the dawn of this new millennium.

Chapter 20 describes the inherent organizational design and operating principles of this new Web as those of living systems, and thus shows why it has the power to force corporations with organizational designs and operating principles based on command and control mechanics to change their ways—to become more like living systems themselves. As corporations, which play such a powerful determining role in our species' behavior as a whole, are transformed into abiding by the sustainable survival principles of living systems, their goals will necessarily come into harmony with our personal and community goals. We can then mature like other species from competition to cooperation and build a human society in which the goals of individual *and* community, of local *and* global economy, of economy *and* ecology are met. This will shift us out of crises and into the happier, healthier world of which we all dream. Let it be so!

Elisabet Sahtouris, September, 2000

1

A Twice-Told Tale

Everyone knows that humanity is in crisis, politically, economically, spiritually, ecologically; any way you look at it. Many see humanity as close to suicide by way of our own technology; many others see humans as deserving God's or nature's wrath in retribution for our sins. However we see it, we are deeply afraid that we may not survive much longer. Yet our urge to survival is the strongest urge we have, and we do not cease our search for solutions in the midst of crisis.

The proposal made in this book is that we see ourselves in the context of our planet's biological evolution, as a still new, experimental species with developmental stages that parallel the stages of our individual development. From this perspective, humanity is now in adolescent crisis and, just because of that, stands on the brink of maturity in a position to achieve true humanity in the full meaning of that word. Like an adolescent in trouble, we have tended to let our focus on the crisis itself or on our frantic search for particular political, economic, scientific, or spiritual solutions depress us and blind us to the larger picture, to avenues of real assistance. If we humbly seek help instead from the nature that spawned us, we will find biological clues to solving all our biggest problems at once. We will see how to make the healthy transition into maturity.

Some of these biological clues are with us daily, all our lives, in our own bodies; others can be found in various ages and stages of the larger living entity of which we are part—our planet Earth. Once we see these clues, we will wonder how we could have failed to find them for so long.

The reason we *have* missed them is that we have not understood ourselves as living beings within a larger being, in the same sense that our cells are part of each of us.

Our intellectual heritage for thousands of years, most strongly developed in the past few hundred years of science, has been to see ourselves as separate from the rest of nature, to convince ourselves we see it objectively—at a distance from ourselves—and to perceive, or at least model it, as a vast mechanism.

This objective mechanical worldview was founded in ancient Greece when philosophers divided into two schools of thought about the world. One school held that all nature, including humans, was alive and self-creative, ever making order from disorder. The other held that the 'real' world could be known only through pure reason, not through direct experience, and was God's geometric creation, permanently mechanical and perfect behind our illusion of its disorder.

This mechanical/religious worldview superseded the older one of living nature to become the foundation of the whole Western worldview up to the present.

Philosophers such as Pythagoras, Parmenides, and Plato were thus the founding fathers of our mechanical worldview, though Galileo, Descartes, and other men of the Renaissance translated it into the scientific and technological enterprise that has dominated human experience ever since.

What if things had gone the other way? What if Thales, Anaximander, and Heraclitus, the organic philosophers who saw all the cosmos as alive, had won the day back in that ancient Greek debate?

What if Galileo, as he experimented with both telescope and microscope, had used the latter to seek evidence for Anaximander's theory of biological evolution here on Earth, rather than looking to the skies for confirmation of Aristarchus's celestial mechanics? In other words, what if modern science and our view of human society had evolved from organic biology rather than from mechanical physics?

We will never know how the course of human events would have differed had they taken this path, had physics developed in the shadow of biology rather than the other way around.

Yet it seems we were destined to find the biological path eventually, as the mechanical worldview we have lived with so long is now giving way to an organic view—in all fairness, an organic view made possible by the very technology born of our mechanical view.

The same technology that permits us to reach out into space has permitted us to begin seeing the real nature of our own planet to discover that it is alive and that it is the only live planet circling our Sun.

O   O   O

The implications of this discovery are enormous, and we have hardly even begun to pursue them. We were awed by astronauts' reports that the Earth looked from space like a living being, and were ourselves struck by its apparently live beauty when the visual images were before our eyes. But it has taken time to accumulate scientific evidence that the Earth is a live planet rather than a planet with life upon it, and many scientists continue to resist the new conception because of its profound implications for change in all branches of science, not to mention all society.

The difference between a planet with life on it and a living planet is hard at first to understand. Take for example the word, the concept, the practice of *ecology*, which has become familiar to us all within just the few short decades that we have been aware of our pollution and destruction of the environment on which our own lives depend.

Our ecological understanding and practice has been a big, important step in understanding our relationship to our environment and to other species. Yet, even in our serious environmental concern, we still fall short of recognizing ourselves as part of a much larger living entity. It is one thing to be careful with our environment so it will last and remain benign; it is quite another to know deeply that our environment, like ourselves, is part of a living planet.

The earliest microbes into which the materials of the Earth's crust transformed themselves created their own environments, and these environments in turn shaped the fate of later species, much as cells create their surround and are created by it in our own embryological development.

As for physiology, we already know that the Earth regulates its temperature as well as any of its warm-blooded creatures, such that it stays within bounds that are healthy for life despite the Sun's steadily increasing heat. And just as our bodies continually renew and adjust the balance of chemicals in our skin and blood, our bones and other tissues, so does the Earth continually renew and adjust the balance of chemicals in its atmosphere, seas, and soils. How these physiological systems work is now partly known, partly still to be discovered, as is also still the case with our bodies' physiological systems.

Certainly it is ever more obvious that we are not studying the mechanical nature of Spaceship Earth but the self-creative, self-maintaining physiology of a live planet.

Many still take the live Earth concept, named Gaia after the Earth goddess of early Greek myth, more as a poetic or spiritual metaphor than as a scientific reality. However, the name Gaia was never intended to suggest that the Earth is a female being, the reincarnation of the Great Goddess or Mother Nature herself, nor to start a new religion (though it would hardly hurt us to worship our planet as the greater Being whose existence we have intuited from time immemorial). It was intended simply to designate the concept of a live Earth, in contrast to an Earth with life upon it.

Actually, Gaia, or the Roman form, Gea, was an earlier name for our planet than Earth. It was lost in the wandering of words from ancient Greek through other languages to English. In Greek, our planet has always been called Gaia in its alternate spelling *Ge*, which we see in English words taken directly from Greek, such as *geology*, the formation of the Earth; *geometry*, the measurement of the Earth; and *geography*, the mapping of the Earth. In accord with our own practice of calling planets by the names of Greek deities in their Roman versions, we really should call the Earth Gea. Greek, like English, has always used the same word for Earth-as-world and Earth-as-ground—the ancient Ge that became the modern Gi, pronounced Yee. The English word Earth came from an ancient Greek root meaning working the ground, or earth—*ergaze*—which evolved into the name of the Nordic Earth goddess, Erda and then into the German Erde and the English Earth. Thus even the word Earth implies a female deity.

With that digression intended to make the name Gaia more acceptable to those who still consider the name and image somehow inappropriate for a scientific concept, let us look also at the myth itself—the creation myth of Gaia's dance.

The story of Gaia's dance begins with an image of swirling mist in the black nothingness called Chaos by the ancient Greeks—an image reminding us of modern photos of galaxies swirling in space. In the myth it is the dancing goddess Gaia, swathed in white veils as she whirls through the darkness. As she becomes visible and her dance grows ever more lively, her body forms itself into mountains and valleys; then sweat pours from her to pool into seas, and finally her flying arms stir up a windy sky she calls *Ouranos*—still the Greek word for sky—which she wraps around herself as protector and mate.

Though she later banishes Ouranos—Uranus, in Latin—to her depths for claiming credit for creation, their fertile union as Earth and Heaven brings forth forests and creatures including the giant Titans in human form, who in turn give rise to the gods and goddesses and finally to mortal humans.

From the start, says the myth—true to human psychology—people were curious to know how all this had happened and what the future would bring. To satisfy their curiosity, Gaia let her knowledge and wisdom leak from cracks in the Earth at places such as Delphi where her priestesses interpreted it for people.

Our curiosity is still with us thousands of years after this myth served as explanation of the world's creation. And in a sense, Gaia's knowledge and wisdom are still leaking from her body—not just at Delphi, but everywhere we care to look in a scientific study of our living planet.

The new scientific story of Gaian creation has other parallels to the ancient myth. We now recognize the Earth as a single self-creating being that came alive in its whirling dance through space, its crust transforming itself into mountains and valleys, the hot moisture pouring from its body to form seas. As its crust became ever more lively with bacteria, it created its own atmosphere, and the advent of sexual partnership finally did produce the larger life forms—the trees and animals and people.

The tale of Gaia's dance is thus being retold as we piece together the scientific details of our planet's dance of life. And in its context, the evolution of our own species takes on new meaning in relation to the whole. Once we truly grasp the scientific reality of our living planet and its physiology, our entire worldview and practice are bound to change profoundly, revealing the way to solving what now appear to be our greatest and most insoluble problems.

From a Gaian point of view, we humans are an experiment—a young trial species still at odds with ourselves and other species, still not having learned to balance our own dance within that of our whole planet. Unlike most other species, we are not biologically programmed to know what to do; rather, we are an experiment in free choice.

This leaves us with enormous potential, powerful egotism, and tremendous anxiety—a syndrome that is recognizably adolescent.

Human history may seem very long to us as we study all that has happened in it, but we know only a few thousand years of it and have existed as humans for only a few million years, while Earth has been self-creating and evolving for billions of years. We have scarcely had time to come out of species childhood, yet our social evolution has changed us so fast that we have leaped into our adolescence.

Humans are not the first creatures to make problems for themselves and for the whole Gaian system, as we will see. We are, however—unless whales and dolphins beat us to it in past ages—the first Gaian creatures who can understand such problems, think about them, and solve them by free choice. In fact, the argument of this book is that our maturity as a species depends on our accepting the responsibility for our natural heritage of behavioral freedom by working consciously and cooperatively toward our own health along with that of our planet.

Our ability to be objective, to see ourselves as the *I* or *eye* of our cosmos, as beings independent of nature, has inflated our egos—*ego* being the Greek word for *I*. We came to separate the *I* from the *it* and to believe that 'it'—the world apart from us, out *there*—was ours to do with as we pleased. We told ourselves we were either God's favored children or the smartest and most powerful naturally evolved creatures on Earth. This egotistic attitude has been very much a factor in bringing us to adolescent crisis.

And so an attitude of greater humility and willingness to accept some guidance from our parent planet will be an important factor in reaching our species maturity.

The tremendous problems confronting us now—the inequality of hunger on one side and overconsumption on the other, the possibly irreversible damage to the natural world we depend on, just as our cells depend on the wholeness of our bodies for *their* life—are all of our own making. These problems have become so enormous that many of us believe we will not be able to solve them in time.

Yet just at this time in our troubled world we stand on the brink of maturity, in a position to recognize that we are neither perfect nor omnipotent, but that we can learn a great deal from a parent planet that is also not perfect or omnipotent but has the experience of billions of years of overcoming an endless array of difficulties, small and great.

When we look anew at evolution, we see not only that other species have been as troublesome as ours, but that many a fiercely competitive situation resolved itself in a cooperative scheme. The kind of cells our bodies are made of, for example, began with the same kind of exploitation among bacteria that characterizes our historic human imperialism, as we will see.

In fact, those ancient bacteria invented technologies of energy production, transportation and communications, including a WorldWideWeb still in existence today, during their competitive phase and then used those very technologies to bind themselves into the cooperative ventures that made our own existence possible. In the same way, we are now using essentially the same technologies, in our own invented versions, to unite ourselves into a single body of humanity that may make yet another new step in Earth's evolution possible. If we look to the lessons of evolution, we will gain hope that the newly forming worldwide body of humanity may also learn to adopt cooperation in favor of competition. The necessary systems have already been invented and developed; we lack only the understanding, motive, and will to use them consciously in achieving a cooperative species' maturity.

It may come as a surprise that nature has something to teach us about cooperative economics and politics. Sociobiologists, who have told us much in recent decades about humanity's animal heritage, have tended to paint us a bleak picture. Calling on our evolutionary heritage as evidence that we will never cure ourselves of territorial lust and aggression toward one another, they continue to predict there will be no end to economic greed and political warfare. But it is the aim of this book to show that these sociobiologists have presented a misleading picture—as misleading as earlier scientists' one-sided view of all natural evolution as "red in tooth and claw," the hard and competitive struggle among individuals on which we have modeled our modern societies.

The new view of our Gaian Earth in evolution shows, on the contrary, an intricate web of cooperative mutual dependency, the evolution of one scheme after another that harmonizes conflicting interests.

The patterns of evolution show us the creative maintenance of life in all its complexity. Indeed nature is more suggestive of a mother juggling resources to ensure each family member's welfare as she works out differences of interest to make the whole family a cooperative venture, than of a rational engineer designing perfect machinery that obeys unchangeable laws.

For scientists who shudder at such *anthropomorphism*—defined as reading human attributes into nature—let us not forget that *mechanomorphism—reading mechanical attributes into nature*—is really no better than second-hand anthropomorphism, since mechanisms are human products. Is it not more likely that nature in essence resembles one of its own creatures than that it resembles in essence the nonliving product of one of its creatures?

The leading philosophers of our day recognize that the very foundations of our knowledge are quaking—that our understanding of nature as machinery can no longer be upheld. But those who cling to the old understanding seriously fear that all human life will break down without a firm foundation for our knowledge of nature in mathematical reference points and laws of physics. They fail to see what every child can see—that humming-birds and flowers work, that nature does very well in ignorance of human conceptions of how it must work.

Machinery is in fact the very antithesis of life. One must always hope a machine, between its times of use, will not change, for only if it does not change will it continue to be of use. Left to its own devices, so to speak, it will eventually be destroyed by its environment. Living organisms, on the other hand, cannot stay the same *without* changing constantly, and they use their environment to their advantage. To be sure, our machinery is getting better and better at imitating life; if this were not so, a mechanical science could not have advanced in understanding. But mechanical models of life continue to miss its essential self-creativity. Fortunately, our survival struggle is leading to intuitive grasps of nature's principles that are shifting our technologies into serving cooperative life pur-poses, especially clearly in the phenomenon of the global Internet.

O   O   O

We are learning that there is more than one way to organize functional systems, to produce order and balance; that the imper-fect and flexible principles of nature lead to greater stability and resilience in natural systems than we have produced in ours—both technological and social—by following the mechanical laws we assumed were natural.

We designed our societies as though they were machinery; we made a Cold War on one another over who had the perfect social design. Our greatest recent conflict was over whether individuals should sacrifice their individual interest to the welfare of the whole or whether individual interest should reign supreme in the hope that the interests of the whole would thus take care of themselves.

No being in nature, outside our own species, is ever confronted with such a choice, and if we consult nature, the reason is obvious. The choice makes no sense, for neither alternative can work. No being in nature can ever be completely independent, although independence calls to every living being, whether it is a cell, a creature, a society, a species, or a whole ecosystem.

Every being is part of some larger being, and as such its self-interest must be tempered by the interests of the larger being to which it belongs. Thus mutual consistency works itself out everywhere in nature, as we will see again and again in this book.

For clues on organizing a workable economics and politics, we need not even look beyond our own bodies, with their cooperative diversity of cells and organs as a splendid example to us in working out our social future.

Diversity is crucial to nature, yet we humans seem desperately eager to eliminate it, in nature and in one another. This is one of the greatest mistakes we are making. We reduce complex ecosystems to one-crop monocultures, and we do everything in our power to persuade or force others to adopt our languages, our customs, our social structures, instead of respecting their diversity and recognizing its validity. Both practices impoverish and weaken us within the Gaian system.

We are right to worry about our survival, for we foolishly jeopardize it.

We are wrong to devote our attention to saving or managing nature. Gaia will save herself *with or without us* and hardly needs advice or help in managing her affairs. To look out for ourselves, we would be wise to interfere as little as possible in her ways, and to learn as much as possible of them.

Our technology has ravaged nature and continues to do so, but the ravages of technology are rooted in our youthful species' greed, our single bottom-line quest for profits motive. There is no intrinsic reason that we humans cannot develop a benign technology once we agree that our desire to maximize profits is completely at odds with nature's dynamic balance—that greed prevents health and welfare for all. As Janine Benyus has pointed out, we assigned one group of people called biologists to study how other species make their living, and a completely separate group of people called economists to determine how *our* species makes its living.

No other creatures take more than they need, and this must be our first lesson. Our second lesson is to learn and emulate nature's fine-tuned recycling economics, largely powered by free solar energy. This does not mean going back to log cabins or tipis, but to eliminate waste and junk as we creatively develop diverse human lifestyles of elegant and sustainable simplicity.

The purpose of this book is to help pave the way to a happier and healthier future through an understanding of our relationship to the Gaian Earth system that spawned us and of which we are part—a great being that, however it may annoy us, is not ours to dominate and control. We can damage it, but we cannot run it; we had better try to find out what it is all about and what we *are* doing, and *may* do, to survive happily within it.

The aggressive and destructive motives of domination, conquest, control, and profit have been presented to us as unchangeable human nature by historians as well as by sociologists. But mounting evidence from archaeology strongly suggests that human societies were, for the greater part of civilized history, based more on cooperation and reverence for life and nature than on competition and obsession with death and technology. It seems our human childhood, which lasted far longer than has our recent adolescence, was guided by religious images of a near and nurturing Mother Goddess before a cruel and distant Father God replaced her in influence. As we come out of adolescence we often recognize the value of what we were taught in childhood, and this new historical view of ourselves supports the general thesis of this book.

Like Gaian creation itself, human understanding or knowledge ever evolves.

Parts of the story you are about to read will already have changed by the time you read it. Others will change in the years to come as new things about Earth-Gaia and about human history are discovered. Any of us is free to help find new pieces of the story, bring those we know up to date, and then reinterpret the evidence as a whole, for in the last analysis, every interpretation has its personal color and flavor.

The next chapter is concerned with cosmic beginnings as a living context for our living planet; succeeding chapters, up to half of this book, tell of Gaian evolution over billions of years before we humans become part of it. Those interested in the story of human society may be tempted to skip this part of the story, but the scientific account of evolution in this book is not separable from our human social history. The details of our biological heritage from ancient bacteria on are given because therein lie the clues to a better human future. It is only within this context that we can appreciate our newness and our differences from the rest of nature, to see at the same time how we can benefit from its vast experience to fit ourselves in more harmoniously.

It is on this that everything now depends; species suicide is our only alternative, and there is really no reason to make a dramatic adolescent exit instead of growing up, taking on adult responsibility, and reaping the pleasures of productive maturity. Let us then follow the evolution of Gaian creation and of our own history as social and technological creatures within this great dance of life. Let's see what meaning and guidance all this may give in our present crisis, to speed us on our way into full maturity, to a happier future in which we promote our own health and that of our planet within the greater cosmic dance.

2

Cosmic Beginnings

The Greek myth of Gaian creation began with an image of the goddess whirling out of darkness, wrapped in flowing white veils. In ancient India the very beginning of the universe, or cosmos, was imagined as swirls in a sea of milk.

We will probably never know how ancient peoples understood that the first forms to create themselves were whirling white spirals. However they knew, we in our own day can actually see just what those first swirling white forms out in space really were. We call them protogalaxies, or first galaxies. And we have learned that whole protogalaxies do dance as whirling white forms in space long before planets evolve within them, and longer before creatures can evolve as parts of planets.

The material universe, as most scientists describe it today, began with a huge explosion of energy they call the Big Bang. Some say this explosion was more like a great wave of energy rising out of an even greater sea of energy; others talk about continuous creation as well as an initial event; some of those tell us matter is continually

created from an underlying intelligent source, such as consciousness. Whatever happened to start our universe, our current scientific story is that it began as very hot, explosively fast-moving
energy that has been spreading and cooling ever since, creating
spacetime as it does so.

The ancient Greek word *chaos* first denoted nothingness—the
great void before there was anything material in the universe.
(They also spoke of a fullness of potential named the *plenum*.)
Later, chaos came to mean anything so mixed up or messed up that
it has no pattern, no order, no meaning, at least none that we
humans can detect. (The word *random* carries the same meaning of
lack of order or pattern.) With chaos theory, we began to see chaos
as having hidden pattern—pattern we are unable to detect. All
these ways of using the word chaos have been used to describe the
beginning of the universe. There was nothingness, as no-thing had
been formed; yet the dance of energy that would create order or
pattern had begun.

The word *cosmos* was coined as the opposite of chaos, to mean
order as opposed to disorder, form and pattern instead of formless-
ness and lack of pattern, things instead of no-things, a world
instead of no world. The first Greek philosophers understood creation as a process of turning disorderly or non-orderly chaos into
an orderly cosmos, and we have no better way of describing it
today. For as the chaotic hot energy cooled and spread, it turned
itself into a great dance of spiraling cosmic patterns

Our best explanation of how this happened begins with the
idea of imbalance, as it also did in many ancient philosophies. In
the early chaos, as the explosive energy spread and cooled, there
must have been pockets of more or less energy, or, as energy
formed itself into particles, pockets of more or fewer particles, or

different numbers of different kinds of particles. Any such imbalances would have set up currents of motion among the heavier slower-moving particles in the overall force of out-thrusting universal energy.

Particles, or subatomic particles, are the tiniest whirling packets of pure energy from which all matter—all the stuff of the universe—is made. The whirling energy of particles created a new force, or forces, among particles, so that when early cosmic particles passed close enough to each other to attract each other, some of them held together as simple atoms. We can imagine this as rather like people dancing, attracting each other when close enough to whirl each other about. Other particles were pushed apart, while most particles kept zooming along alone among the first slower atoms of floating gas.

The physical force that still works at the greatest distance among the clumps of matter that formed in our universe is the one we call gravitation; two others—the strong and weak nuclear forces—have their effect inside atoms and stars. The fourth and last to develop was the electrical force, which works to combine atoms into molecules, but that is getting ahead of our story. Some new theories describe gravitation as a basic property of the zero-point energy field, rather than as a force. It is wise to note that our theories are still evolving rapidly and that this story may still change dramatically.

Natural, or physical, influences, then, on great and small levels, pulled and pushed the universe into patterns great and small. As the number of atoms, and the explosive young universe itself, grew larger, imbalances here and there drifted and swirled the atoms into great gas clouds. These clouds formed more swirls within themselves, some of the thickest becoming protogalaxies sparking with light.

Light is made of energy packets we call photons. New photons can be created like tiny sparks when other fast-flying particles bump into one another very hard. Photons make the protogalaxies visible, and it now seems they are created continually everywhere in the universe, even inside *us*.

If an ancient storyteller could have looked through a modern telescope to see a protogalaxy forming, he might well have said, "Ah, you see, there is the white-veiled Gaia whirling about in her dance." A modern scientist, on the other hand, sees such protogalaxies as the natural result of imbalances and forces in the great cosmic energy field—a swirling of disorderly or chaotic matter into orderly or cosmic patterns, a sea of energy whose forceful currents form natural whirlpools large and small. This is especially important to recognize: that the largest patterns—the great swirling clouds within which protogalaxies took shape—were forming almost as soon as the tiniest particles and atoms began whirling into being. Our universe, or cosmos, has always been a dance of interactions among the large and small moving patterns, each contributing to the other's formation. It was not built from the top down *or* from the bottom up, but evolved as a dance between great and small.

But can we really see protogalaxies forming billions of years ago while looking through telescopes now? Is it possible to look back into time, and so very far back at that?

We can. With modern telescopes we can see back to nearly the beginning of the universe! Magical as it seems, the explanation for this strange power we have is quite simple. Everything we see comes to our eyes as light photons that have bounced off or come out of whatever we are looking at. Light bounces off a cat or a cloud, for instance, and comes out of a candle flame or a star. But what exactly is light?

We've already talked about photons as energy particles created when other particles bump into one another. Stars and flames are made of atoms and particles moving so fast that unusual numbers of photons are created in them.

Photons travel through space in waves of different lengths and strengths, some of which we see as different colors and brightnesses when they get to our eyes. Though light is extremely fast by human standards—at 186,000 miles per second—it still takes some time to get from an object that created it, or from one it has bounced off, to our eyes. The time it takes light to travel holds the secret of looking back in time.

It takes about seven minutes for light to get from the Sun to our eyes. Every time we look at the Sun, we are seeing the light pattern that left it seven minutes ago. That means we are seeing the Sun the way it was seven minutes ago and not as it is the moment we are looking at it. The Sun is the star nearest to us. Other stars are so far away that their light takes years to get to our eyes—thousands of years, even millions of years, depending on how far away the star is. The distance of stars, in fact, is measured in light-years—the number of years it takes for their light to reach us.

Whenever you look up at the night sky, even without a telescope, you are looking back into time. You see each star as it was when the light reaching your eyes left it. By looking at many stars, you are looking at many times past. How far past depends, of course, on the distance of each star. The farther away the star is, the longer ago it sent out the information about what it looks like—that is, the light pattern of the star that has finally found your eyes.

Our own galaxy, the Milky Way, is shaped like a giant swirling pinwheel within an enormous but less visible spherical torus. It takes light a hundred thousand years to cross it. If there are any creatures on another planet—say, three thousand light years from us, in our own galaxy—who are looking at us right now, what do they see? If their telescopes are powerful enough, they may be seeing a storyteller speaking of Gaia's dance in an ancient Greek village!

O   O   O

Powerful telescopes can pick up light that is too weak from its long travels for our eyes alone to see—even light from stars and galaxies so old that they were among the first stars and galaxies, or protogalaxies, in the universe, so old they are just beginning cosmic creation. Let's watch one of them in its evolving dance.

Inside the spiraling veils of hydrogen gas, which is made of the first and simplest atoms in the universe, smaller rolling waves create a ring of denser atoms, of more intense energy, at the center. Around it, great loose balls of gas form, something like the way dust balls form under a bed. In the center of such balls, the lively atoms and particles are pulled ever closer together by physical forces until it gets very hot from all the crowding. As these gas balls get hotter and heavier, they become stars.

Wherever we look back into ancient skies, we see galaxies taking shape and growing through different stages. Inside the first generation of stars the incredible heat and pressure begins causing what we now know as nuclear reactions—the transformation of one kind of atom into another. The first such reaction squeezes hydrogen atoms together to form helium atoms, which is what our Sun is doing all the time. This process creates heat and light, some of which escapes from the stars in spreading waves of photons. The burning gases on the outside of stars pull away in waves, like the skin a snake sheds, because of the gravitational pull of matter, such as other stars, around them. Stars must constantly keep their balance between tremendous forces pulling them apart and other forces squeezing them together.

Eventually, the first-generation stars collapse from growing so heavy they can no longer keep their balance between the internal and external pulling. Their atoms mass ever more tightly together. Eventually the star implodes and then explodes, scattering stardust like seeds back into the galactic gas cloud. The mother cloud becomes ever thicker with the gas and dust of such explosions and gives birth to a new generation of stars as the old ones die.

The next generation of stars forges its atoms into yet bigger and heavier kinds until all the different kinds of atoms—all the different elements of the universe—have been formed from the original hydrogen atoms. Meanwhile, the central ring of gas clouds in a galaxy grows larger and more complicated, becoming a kind of skeleton that holds the galaxy together. At last many of the atoms from exploding stars are too heavy to form new stars and begin to form themselves into planets circling around stars that are made of the lightest elements. This is why our Sun, although it is not a first-generation star, is made like one. The heavier elements of its parent star are in its planets.

So protogalaxies evolve into galaxies—whirling, weaving, squeezing, exploding, pulsing their insides into ever richer patterns and parts. Molecules formed of groups of atoms, even the kinds of molecules from which the familiar living systems of the Earth formed themselves, are created in complex galactic processes, as we shall see later. For now, let us remember that the stars we see in our night skies are only a few of those in our own galaxy, and, as we see them with our eyes, they don't begin to hint of our galaxy's complex patterns and processes. Far beyond those stars lie billions of other galaxies, each made of billions of stars and planets wheeling in their clouds of gas and dust, creating who knows how much life.

Astronomers, whose name comes from the Greek word for star, *astron,* now know the different shapes of individual galaxies and can see them clustered into larger patterns. There are even clusters of clusters, called superclusters, even some greater pattern that extends all through the universe, parts of it appearing in the images we have been able to make of them, like huge curved strings and the holes in Swiss cheese. These still crude images, we may hope, will one day resolve themselves into an understanding of the greatest patterns of all.

Though we don't know what these patterns are as yet, it appears increasingly obvious that they form a cosmic unity of process and pattern rather than a chaotic spray of unrelated parts. A single notion that would account for such pattern is the concept of *mutual consistency,* which is at the heart of 'bootstrap philosophy,' a mathematical physics conception popularized by Fritjof Capra. This is the concept that the universe is a dynamic web of events in which no part or event is fundamental to the others since each follows from all the others, the relations among them determining the entire cosmic pattern or web of events. In this conception, all possible patterns of cosmic matter-energy will form, but only those working out their consistency with surrounding patterns will last.

Mutual means shared; consistency means agreement or harmony. Thus we can sense mutual consistency as the shared harmony worked out among cosmic patterns. The notion can be made more familiar by considering the shared social harmony worked out by groups of people when each individual adjusts his or her behavior to that of the others in a harmonious way. Anyone who cannot do this will tend to be excluded from the group, unless the deviant can force the others to make their behavior consistent with his or hers, in which case a new (if tenuous) mutual consistency would have been worked out. At present our species is not behaving in a way that is mutually consistent with the other species and features of our planet, and the consequences may preclude our survival.

Increasingly, then, we are discovering with modern instruments and measurements what ancient peoples told in myth—that all of the universe is one great pattern, a single dance evolving into ever richer complexity over billions of years.

Until recently, scientists had a rather different idea of how nature forms itself—a mechanical idea of wholes built from parts as machinery is built, though coming together automatically without any designer or builder. We shall learn more of this way of looking at things later, when we look at human history. For now what matters is to understand this new way of seeing that all evolution—of the great cosmos and of our own planet within it—is an endless dance of wholes that separate themselves into parts and parts that join into mutually consistent new wholes. We can see it as a repeating, sequentially spiraling pattern: unity—>individuation—>competition—>conflict—>negotiation—>resolution—>cooperation—>new levels of unity, and so on.

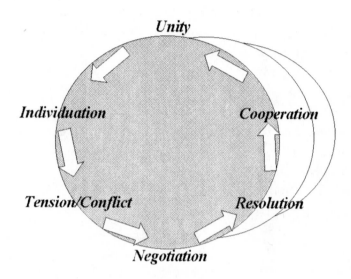

Evolution's Repeating Cycle

We have already seen how the swirling gas clouds that evolved into galactic clusters began forming as soon as particles joined together to form the first simple hydrogen atoms. The early universe thus evolved by forming more and more parts within itself, many of them becoming new wholes in their own right if they proved consistent with other wholes surrounding them. As stars form within a protogalaxy, it becomes a galaxy—a great star system that in turn forms within itself relatively independent single or double star systems, some with planets such as our solar system. Later we will see how a planet's crust can form the packets of life we call microbes, or bacteria, and how these in turn can join together in building larger living cells, which in their turn evolve into larger creatures.

The universe of all these parts within parts, or wholes within wholes, reminds us of nesting boxes or of the Chinese or Russian dolls of various sizes that fit inside one another. The philosopher scientist Arthur Koestler suggested we call each whole thing within nature a *holon*—a whole made of its own parts, yet itself part of a larger whole. A universe of such holons within holons is, then a *holarchy*—in Greek, a source of wholes—one original whole that formed ever more complicated smaller wholes within itself, some becoming holarchies themselves. We will use this image and the terms *holon* and *holarchy* throughout this book to show the embeddedness of natural entities.

Holons in Holarchy

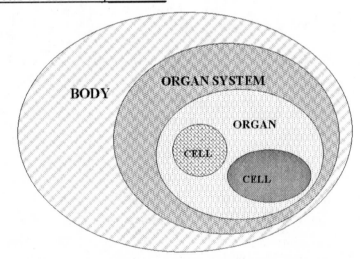

Our own solar system, with its Sun-star nucleus surrounded by planets, Moons, asteroids, comets, and space dust, is a holon within the larger holon of our galaxy. It was born of the scattered gases and stardust of an older star that became a supernova exploding about five billion years ago, maybe even more than one of them. The Earth is still so radioactive from this explosion that its core is kept hot by continuing nuclear reactions, and many atoms all over its surface—in rocks and trees and even in our own bodies—are still exploding.

In our bodies it has been estimated that three million potassium atoms explode every minute. These explosions are much too tiny for us to see, feel, or perceive in any other way. They are not arranged to blow up neighboring atoms as well as themselves, as in our powerful man-made nuclear chain reactions. Still, they are evidence that stardust is not just fairy-tale magic; it is what we are really made of—we and everything else that is part of our world.

Between five and four and a half billion years ago, some of the gas and dust from that great star explosion gathered into an Earth-ball made of twelve different kinds of atoms, or twelve elements. As it condensed, it grew heavier and spun around faster. The heat of pressure and nuclear reactions inside it melted the packed matter into a fiercely burning liquid. But the outside of this fiery ball, touching cold space, cooled off as a thin crusty skin, a bit the way homemade pudding forms a skin as it cools, or the way fat hardens on top of cooling gravy. The Earth's skin was made of rock—a crust of rock around a hot, molten mantle of magma, with its heaviest elements at its solidifying core.

While it was still very thin, this crust melted again and again, each time letting the heaviest metal elements sink back towards the core while lighter elements formed a foam of rock around those fiery insides. Today's Earth has a thicker crust, broken up into great *tectonic plates* that ride on the denser mantle surrounding the solid core. We can still see the hot liquefied elements of the mantle pouring out through volcanoes puncturing the crust. And in Earthquakes we can feel the motion of the great tectonic plates as they slide about creating new geological formations.

In the myth of Gaia's dance, as her body forms mountains and valleys, the seas are formed from her warm moisture. Just so, it seems, the seas eventually pooled on the young Earth.

At first, when the Earth's crust cracked here and there, the liquid magma insides oozed out as lava. Lava, as the pressure that keeps it together is released, separates into heavy atoms that cool into more crusty rock, into water that hisses up as steam, and into other atoms light enough to float over or off the surface of the planet as gases. We now believe the water steaming off the hot crust stayed high above an early atmosphere of poisonous (from our point of view) gases for what may have been a long time, but eventually formed clouds that condensed into rain. The rain poured down so hard and for so long that the seas began pooling on top of the heavier rock. As more and more lava oozed through cracks in the Earth's crust, the crust itself grew thicker and lumpier; as new clouds gathered and fell in cycles, the seas grew ever bigger and deeper.

As the Earth's crust grew thicker, new streams of lava broke through it with greater force. Spitting volcanoes shot their fiery insides high into the air, forming mountains as the lava cooled and hot ashes settled down. More mountains were formed when Earthquakes cracked the crust and slid parts of it over one another, and when the crust heaved and bulged without breaking. Rocks sliding over one another were ground into sand and dust.

Huge dust clouds were created when meteors of all sizes—some of them as large as small planets—struck the Earth, smashing into the crust, pitting it, breaking it up, mixing it with the space rocks themselves.

The gases floating around the planet, those just heavy enough to be held by its gravity, were nothing like the air we breathe now. There was no oxygen, but only a mixture of gases which, had the Earth not come alive, would have eventually settled into something like the atmosphere on Venus and Mars today—an atmosphere without oxygen around a lifeless planet.

What, then, did the Earth have that Venus and Mars did not? James Lovelock, author of the Gaia hypothesis called one of its special features the 'Goldilocks effect:' Venus was too hot, Mars was too cold, but the Earth was just the right temperature for life. Another was its water, enough of it in liquid form, in this just-right temperature, to carry supplies from place to place as blood is carried through a body. The constant transport of supplies must be possible for life to evolve.

Everything of Earth's surface—oceans and rivers, mountains and fertile fields, forests and flowers, creatures that float or fly or crawl or climb, everything, including ourselves, is actually made from the same original but recycled supplies, except for the small input of meteors. Our world has created itself as new arrangements of the same atoms that started out inside a star, then formed the molten metal, crusty rock, and gases of a newborn planet—a planet that covered itself in seas as we have seen and is now ready to go on with its dance of life. Let's follow this great Gaian recycling system to see just how stardust continues to transform itself into a living planet—into all the amazing complexity of our beautiful world.

3

The Young Earth

Shall we think of the young Earth at this point as a lifeless planet on which life is about to evolve? Most of us have been taught in school that animate matter is one thing—it is alive—and that inanimate matter is quite another, for it is not alive. Are the Earth's rocky crust and watery seas inanimate, lifeless matter while the plants and animals we know this story is leading up to are made of animate, or living, matter? Just what do we mean by the word life?

It may surprise you to learn that scientists do not agree on what life is. Some change their minds from time to time; others don't worry about the question "What is life?" believing the answer is known in some other science. In ancient Greece, when philosophers believed that all nature was alive, a physicist was someone who studied nature— *physis*—and so was concerned with living things. Later, when scientists decided to divide the world into animate and inanimate matter, physicists took on the job of describing how inanimate matter is put together, and biologists, whose name comes from *bios*—way of life—took on the job of describing living things.

Physicists think biologists know what life is because it is their job to know, but biologists keep changing their definition of life and they pass the question of how to tell life from non-life on to chemists, whose name comes from ancient roots having to do with the transformation of matter from one kind into another. So chemists divide chemistry up, in their turn, into two kinds: organic chemistry, the study of living matter, and inorganic chemistry, the study of nonliving matter. Chemists know something about the transformation of inorganic matter into organic matter, but the question of just when and where life began on our planet still gets tossed back and forth among them, or taken back to ideas from physics.

Some scientists talk about life in terms of *non-equilibrium thermodynamics.* This contrasts it with the *equilibrium dynamics* of nonliving things—the physicists' way of solving the problem they created long ago when they declared that life was separate from non-life. Whether or not physics is the appropriate branch of science to define life, this new view at least talks about life as a process rather than as a kind of matter, and that seems closer to what life is all about.

Before religion and science parted company, the answer to the question of how life began was easy. Scientists themselves believed that God created living things, such as plants and animals and people, putting them into the nonliving world he had created for them. But later, when scientists tried to explain the world without bringing God into the picture, they were stuck with believing that life is a special kind of matter that somehow comes from lifeless matter. One version of this belief was known as *spontaneous generation*—the belief that worms, for example, sprang from bits of dead garbage or rotting meat.

Louis Pasteur put an end to that, as we are also taught in school. Or did he? His very careful experiments showed that worms come only from eggs, and never directly from garbage. But where did eggs, which are living things, come from? Flies or other insects, also living things. The explanation seemed easy with a theory of evolution: they came from other worms, which had evolved from the smaller, simpler creatures we traced all the way back to microbes— living things so small they can be seen only through microscopes.

But where do microbes come from? That is still difficult to tell, but we assume they come from the simplest molecular systems that could maintain and reproduce themselves. Some biologists believe that life began with small clumps or sacs of organic molecules. The organic molecules themselves are considered nonliving matter that comes alive when they get stuck together in certain ways that permit them to act on each other to form a living system. In other words, scientists *still* believe that life comes from lifeless matter. In this sense, spontaneous generation was not so much disproved as pushed down to things much smaller than dead meat and worms.

We are still stuck with the question of just what life is. What is it that brings the lifeless molecules in some places, on some planets, to life when they are chained and clumped together in certain ways? Even though we are talking about very tiny things, there is still a big jump from nonliving matter to life.

We have already suggested that it might be better to see life as a process than as a kind of matter. Perhaps it would also help if scientists did not keep looking for the answer only in tinier and tinier parts of nature, believing that in doing so they would see just how things are built from the bottom up.

If we begin, instead, by thinking of wholes, or holons, that form their own parts from the top down, so to speak, everything looks very different. Think, for example, of the huge protogalactic cloud holons we talked about in Chapter 2. If we could watch a movie of the evolution of a protogalaxy sped up so that billions of years happened in a few minutes, what would we see? We would see it whirl and throb, grow and change, its parts dissolving and exploding, more complicated new parts forming in their place and even reproducing themselves as the mature galaxy took on its complicated form. Galaxies themselves split apart and merge with others on collision. And within galaxies—perhaps within all of them—some planets produce what we all agree, here on Earth, to be life.

While astronomers may speak of the lives of stars, they do not seriously count stars or galaxies as living beings. Yet galaxies do some of the things by which we all recognize living beings in our everyday experience of Earth, such as keeping their form through many changes within them, creating and replacing their own parts, sometimes even growing and/or dividing to form offspring galaxies.

The most promising definition of life among biologists, in fact, seems very nearly to fit galaxies, if not stars. This is the definition of life we owe to the Chilean biologists Humberto Maturana and Francisco Varela. Their concept of life is a process called *autopoiesis* (pronounced *auto-po-EE-sis*), which in Greek means self-creation or self-production.

An autopoietic unity, or holon, produces the very parts of which it is made and keeps them in working order by constant renewal. An autopoietic holon works by its own rules and creates a boundary that distinguishes it from its environment and through which it exchanges materials with its environment. We do not see such boundaries around galaxies, yet galaxies are visible as distinct entities that maintain their shape while producing and reproducing their parts. The Earth, as we will see, also produces and renews its parts, including the thick atmospheric boundary through which it exchanges radiation energy with its environment.

O   O   O

It seems that as we learn more about our universe, we need to change our scope and the questions we ask about life. Until now we have assumed that the entire universe is nonliving matter except for some matter on planets such as ours. But why should we divide the universe up in this way? Physicists now tell us, as we will discuss further in the last chapter, that the matter-energy of the earliest universe was already, by its very nature, bound to form living systems. Had things been just the tiniest bit different at the beginning, this would not be so and we could not have evolved. Perhaps, then, life evolves as the *essential* process of the cosmos as a whole and is not just something happening at a special point we hunt for in vain.

This is, in fact, becoming an increasingly acceptable hypothesis among physicists who have revived the ancient Greek concept of the source potential, or *plenum,* as a zero-point energy field (ZPF)—the infinite energies existing at *every* point in spacetime and from which source all matter is created. And even beyond that, ever since quantum theory proved so powerful, some physicists have proposed consciousness—a basic universal consciousness—as the source of all creation.

Historically, we see that science took a big step away from religious explanations of the world, and that it is now taking another big step toward a merger with spiritual explanations. The first step involved a shift from seeing the universe as created by an outside intelligence called God, to seeing it as happening solely through the purposeless mechanics of evolved forces and parts. The second step is a shift from mechanical to organic models of nature, with its organics as self-creation process, not blind mechanics. If science 'officially' acknowledges cosmic consciousness to be the continually self-creative source of the material universe, as many individual scientists now do, this step toward an integrated spirit-energy-matter worldview will be completed, while older worldviews, both religious and scientific will fade into history.

O O O

Galaxies are surely a very significant part of cosmic life processes. It certainly seems that our Earth, born from our galaxy, is alive in its own right.

We do not know whether in our own solar system planets such as Mars and Venus began coming to life and then failed to evolve because they could not keep themselves alive. It is ever clearer that, as with the seeds and eggs of plants and animals, far more planets are produced than actually come to life. Planets must have just the right composition and be in just the right relationship to their star to come as alive as has our Earth. Yet even if only a few planets among many succeed in coming to life, there must be billions of living planets in the universe. And the others—the majority of planets that do *not* come alive in their own right—may still play a supporting role in the life of their galaxies.

The creatures we are used to thinking of as alive, such as plants and animals, contain much supporting 'nonliving' matter in their woody trunks and shells and bones, their thorns and hooves and nails, their hair and scales. Nonliving planets may also be very much a part of live galaxies, perhaps even playing important structural roles in their dynamics. What about the Earth itself? Many scientists argue that it cannot be a living being because only its outermost layer—thin as the dewy mist on an apple at dawn—shows signs of life. What, then, we may ask, about a redwood tree, which is ninety-nine percent deadwood with just a thin skin of life on its surface? No one argues that redwoods are not alive.

It is new in modern science to look at the cosmos and the nature of our planet in this way. It is not easy for scientists to jump from seeing the Earth as a nonliving planet that became a home for living creatures, to seeing it as a single living being with its creatures as much a part of it as cells are a part of our bodies. The scientific studies of Earth have been divided, as we said, into studies of living and nonliving matter. Geologists have had the task of explaining how the geological 'mechanisms' of nonliving matter, such as rock, change with time and weathering. Their work was not intended to be mixed up with that of the biologists who study living things, since these living things have been and still are believed by most scientists to arise in ready-made geological environments and either adapt to them or die out.

Now, however, the jobs of geologists and biologists are getting mixed up whether they like it or not, for the same stardust that was transformed into a rocky planet continues to be transformed into living creatures. What we are made of was stardust long ago, transforming itself into rocky Earth crust and, after a long transformative history of evolution, into us.

To make things more complicated, much of the rock that is transformed into live creatures is later transformed back into rock. And so, just as creatures are made of atoms that were once part of rock, almost all rocks on the Earth's surface are made of atoms that were once part of creatures—creatures that built themselves from the atoms of still earlier rocks.

Think about that. The recycling of stardust gets to be a complicated matter as a planet comes to life. Geologists are now just beginning to believe the Russian scientist V. I. Vernadsky, about whom we will say more later, who understood life on Earth as "a disperse of rock"—rock rearranging itself over billions of years; rearranging itself into ever more complicated forms of life from microbes to men.

That alone is enough to mix up geology and biology, but there is even more to it. Our planet never was a ready-made home, or habitat, in which living creatures developed and to which they adapted themselves. For not only does rock rearrange itself into living creatures and back, but living creatures also rearrange rock into habitats—into places comfortable enough for them to live in and multiply.

But let's take it one step at a time and look first at life as rock rearranging itself. How can this happen?

O   O   O

We begin to see that there is more than one way to understand what life is. We just saw it as a mixture of geology and biology. Let's now try looking at it as a mixture of physics and chemistry.

Remember the forces, such as gravitation, that helped create patterns in the cosmic dance of particles and atoms? One of those forces is the electric force that holds atoms together. This force keeps the outer particle dancers of atoms, their electrons, from flying off into space away from their nucleus of heavier particles. This is not entirely unlike the way gravitation keeps planets from flying off into space away from the Sun, though the orbiting electrons are not hard balls like the ones on the old-fashioned atom models that looked like miniature solar systems.

Powerful electrical, or magnetic, fields were set up by the interaction, through the Earth's crust, of the Sun's energy and the molten metal of Earth's core. We might compare this with a giant battery whose energy can be used to do all sorts of work. At the microcosmic level, the electric force allows electrons to dance in two atoms at once, thus holding the atoms together as a molecule. The more atoms that dance together in this way, the larger the molecules formed.

The strong energy of Sunlight coming to the Earth's crust through the thin early atmosphere stirred up the molecular electric force within the great electric fields, creating storms above and breaking up molecules in rock dust, mud, and seawater near deep ocean rifts below, re-forming them into new and larger molecules. When molecules break up and recombine in new patterns, we call it a chemical reaction, since chemistry is the study of such transformations in the patterns of molecules. The energy that stirred up the electrical force recombined many molecules of the Earth's crust.

Such chemical reactions also happen elsewhere in our galaxy. The larger organic molecules such as those of sugars, acids, and lipids (fats) that were formed on the young Earth are also formed in large quantities and great variety somewhere in the center of our galaxy and perhaps all over it. Some of them come to Earth by way of meteors. It is even possible that meteors may fertilize those planet 'eggs' which come to life.

Some chemical transformations, as we said, were due to electrical storms created among clouds of cooled steam in the early atmosphere as the Sun's energy heated Earth's surface. Besides helping large molecules to form, these storms drove a water recycling system, collapsing clouds into rain, which fell on land and sea, the water rising again by evaporation and collecting back into clouds.

Rainwater ran over the rocks, creating grooves that over the eons formed riverbeds and valleys, carrying ground sand and dust full of rock salts to the seas. Rivers and streams thus formed as the bloodstream of our embryo planet, carrying the supplies needed to develop or evolve its life. For a live planet needs not only a great deal of energy but also flowing matter such as atmospheric gases and water to move things about. As we will see, planetary life is not something that happens here and there on a planet—it happens to the planet as a whole.

The largest new molecules probably formed in shallow waters with the help of Sunlight and lightning storms, or perhaps with the help of the Earth's internal energy around cracks in the Earth's crust on the sea floor. Even the Sun's drying heat at the water's edge may have played a role in forming large molecules and packaging them.

Large molecules, such as naturally forming sugars and acids, absorbed a lot of electrical energy, which was then useful in speeding up their chemical reactions to form ever larger molecules—giant molecules built from the simpler sugars and acids. Some scientists believe the giant molecules formed as large molecules lined themselves up on molds or templates of clay or other crystal matter that had regular, repeating surface patterns or notches for the molecules to hold on to. Others believe that the production of giant molecules happened only after the earliest molecular life systems were already organized within tiny capsules.

Earthlife may be described as autopoietic—self-creating—holons forming within the great Earth holon. In all its creatures, from its earliest microbes to later organisms, we find carbon, or rather reduced carbon compounds, which are carbon atoms surrounded by hydrogen atoms, playing essential roles. The lively energized carbon of the Earth combined easily with oxygen, nitrogen, sulfur, and phosphorus to form all sorts of organic molecules and substances. In fact, you are made of very little other than these six elements in their rich variety of combinations.

Among the giant molecules formed from smaller ones were *proteins*—long strings of *amino acids*, which are themselves molecules made of various combinations of a dozen or fewer carbon, nitrogen, hydrogen, and oxygen atoms. Other giant molecules, assembling from both acids and sugars, were those we call *ribonucleic acid* (RNA) and *deoxyribonucleic acid* (DNA). DNA may actually have been a later development of early living systems based on RNA. Whatever the exact sequence, DNA and RNA came to work together with proteins as the copying and building system of life.

DNA molecules are long chains of smaller molecules joined into long, twisted zippers. We have discovered that the teeth of these DNA-molecule zippers act as a four-letter code that can be arranged in endlessly different patterns, just as letters of the alphabet can be written into words and sentences and books. Because of this, a DNA molecule can hold information. After all, information is really anything in formation—anything that is in an ordered pattern rather than in chaos. Some scientists argue that whatever is in formation becomes information only when used by a living system; in this book we define information as anything that is in formation. A book, for example, contains information even if no one reads it, as does a solar system even if no one uses it.

Information, if it can be copied, can be a plan—a plan for a new copy, or a code plan for something else. DNA can copy, or replicate, itself, but not without the help of proteins that can unlock DNA zippers. Once unlocked, the DNA unzips itself into two half-zippers. As these float around in a soup of smaller molecules, the teeth of each half—all letters of the DNA code—attract new partners just like those that were opposite them in the closed zipper, because those are the only ones that fit into place.

Presto! We have two zippers where there was one, and the two are exactly alike if no mistakes have been made. The DNA-protein partnership evolved in such a way that while proteins unlocked DNA zippers, they also got DNA to store plans coded for building more protein as well as more of itself. Thus the DNA-protein partnerships as wholes were capable of reproducing themselves.

This is a bit oversimplified, since viruses, our only examples of RNA or DNA coated by protein alone, have to get inside cells where other things are available in order to reproduce. Nevertheless, protein with DNA or RNA, or both DNA and RNA, formed molecular cooperatives that became the basic reproduction system of carbon-based life. This *genetic* system—DNA is composed of sequences we call *genes* (from the same root as *genesis*)—is usually described as one-way, the DNA code strictly determining the production of proteins, which are the main building materials of living holons within the Earth holarchy. But recent evidence indicates that proteins can in turn affect and change the DNA code. We will get back to this form of cooperation in later chapters.

Less than five percent of DNA is composed of the genes, which are blueprints for the specific proteins of which living creatures are composed. The role of the remaining more than ninety-five percent is still largely a mystery. It is as though we know just what kind of bricks or stones, wood, glass, etc. are used in building an elaborate building, but still do not know how to read the architectural blueprint.

At some point early in the Earth's history there were plenty of the sugar and acid molecules that were needed to build the long chain molecules of RNA, DNA, and protein. And so the formation of these cooperative partnerships very likely became inevitable in the Earth's warm wet mud and shallow seawater where molecules could move about freely and bump into one another. Possibly there was a long time when these partnerships could hardly have been told apart from the thick soup of building materials around them.

Some scientists, however, argue that such partnerships really could not have gotten under way until the molecules were enclosed in sacs, or membranes, that held them together with other supply molecules and protected them from being dissolved. The most likely candidates for such sacs are called liposomes, literally meaning fat bodies. Liposomes, so tiny they can be seen only with an electron microscope, form as hollow spheres of lipid—fat— molecules, something like microscopic soap bubbles, whenever lipid molecules find themselves in water. This is because the tails of these lipid molecules are hydrophobic, or water avoiding, swinging quickly away from water, protecting one another from it by turning inward so that their heads form a tight sphere around them. Sometimes a double-layered sphere forms with water inside and outside, the double layer having all the lipid molecule heads on both surfaces, with all tails between the two layers of heads. This is the typical formation of simple cell walls and persists even in the most complex cells today.

If a soup containing liposomes and a variety of large molecules is repeatedly dried out and liquefied again, the liposomes break open and flatten out during dry times and re-form their spheres in wet times, sometimes around large molecules—even as large as DNA and protein molecules—that may become trapped inside them while they are broken open. Such conditions must often have occurred at the edges of early seas. The liposomes themselves then function as a skin, or membrane, which serves the molecules inside it both as a protection from, and as a connection to, the outside world. The membrane permits selective chemical crossings, allowing some kinds of atoms or molecules to come in and other kinds to pass outward through them. This soon makes the inside environment chemically different from that outside. Such an arrangement fosters the development of chemical cycles that are basic to living cells.

However the first cells formed, protein became the main material of which living creatures built themselves, while RNA and DNA stored the plans and made it possible for living things to multiply. Some protein molecules came to play a particularly important role by speeding up what other molecules did—say, by speeding up the chemical reactions that build new protein or copy DNA. We call these special proteins *enzymes*, and their wonderful talent for speeding up the chemical dance is very important to our planet's life. In fact, the presence of enzymes has been suggested as one way of defining the presence of life, and the first enzymes likely occurred as a widespread chemical Earth event, perhaps both outside and inside early cells.

While details are still missing, this is essentially how the solid and molten crust of the Earth began to rearrange itself into living creatures. Some of its material gassed off into atmosphere, part reformed into seas, some broke up and was washed into the seas. With the help of great amounts of energy, larger molecules formed and joined into partnerships, set up chemical cycles in early liposomes, speeded up their own reactions with enzyme activity, reproduced themselves, and through all this established themselves as living, or autopoietic, holons—the earliest creatures in their own right. These creatures dwelt within the larger living holon that had given them life and to which they gave a new kind of life in turn. Thus on the one hand we can say that tiny separate living holons evolved all over the Earth, but on the other hand we can say that the Earth holon was coming ever more alive as it evolved its own autopoiesis through a new kind of self-packaging chemical activity.

O   O   O

From our old point of view we could see the beginnings of life only as a collection of microbes descending from some primeval cell that formed accidentally somewhere on Earth, giving rise to offspring that were forced to adjust or adapt their way of life to it by natural selection, which we will discuss in Chapter 7. This was a logical way to see things when we formed our concept of life from our study of individual creatures small enough for us to see as wholes. In our new way of seeing life as autopoietic systems that may be as large as the Earth or even larger, we can think of Earthlife as a planetary process—as the chemical reactions of the planet's crust speeding up, transforming the crustal matter into a blanket of masses of microbes, which in turn transform more of the crust into their livable home, as we will see in the next chapter. And while all this happens at the microcosmic level, the macrocosmic events of the largely molten, still radioactive planet keep its crust heaving, cracking, and sliding, pushing up mountains, buckling in valleys, changing the shapes and positions of continents amid its deepening seas. All together, this is the self-creating dance of a living planet driven by its Sun and by its own energy.

One way of looking at all this is to see the Earth as having come alive through all sorts of 'border activity.' The crust that stirred to life was the boundary enclosing the Earth and at the same time connecting it to outside energy from the Sun and to new materials coming in as meteors. Then, the first cells seem to have formed specifically at the boundaries separating and connecting the land and the sea, or separating and connecting the inner magma with the crustal surface at volcanic sea floor vents. These cells' own boundaries made their individual lives possible by separating them from and connecting them to their environment. At all levels from great to small, this border activity can be seen as highly creative and cooperative—a lesson we humans, with the boundaries we have created among ourselves, might well take to heart.

Let's stop to imagine that we are watching a fast-running movie of the early Earth as it evolves within the larger being of our Milky Way galaxy. As we approach the Earth, we see it whirling and heaving, its thin crust rising and falling, breaking and slipping, bleeding lava where it tears open and sighing bursts of steam. Meteors and planetoids, which are part of the supernova's debris, strike and wound the Earth, making great splashes of molten rock and gas. The thin atmosphere is often reddish with smog produced by the reactions of its own gases. Lightning flashes, and seas form during heavy rains until masses of land and sea become distinct, though the seas are brownish beneath the murky atmosphere.

Slowly the crust thickens and cracks into plates that slide slowly over the surface, carrying the land masses into new patterns. Patches of colored microbes appear and grow along the shores; gradually a tougher but clearer atmospheric skin develops, making the seas turn a sparkling blue. Meteor impact is low; turmoil subsides, and much of the land becomes covered in green. Now and then ice moves down over the green before withdrawing again to the poles, raising and lowering the level of the seas, covering and uncovering the land as though the whole planet is breathing in some gargantuan rhythm. Everything is in constant motion as the Earth shimmers and glows in the Sun against the darkness of space, its changing cloud patterns swirling over blue seas and varicolored lands.

These changes actually happened over billions of years, at a rate too slow for us to recognize as very active. Yet a billion years to our planet is less than a decade is to us. When we use our imagination to see these changes within the time span of a short film, the truly amazing thing is that our planet looks very much like a living creature—perhaps the great cell that popular science writer Lewis Thomas saw it as.

Our movie makes the young planet appear to be trying hard to express itself in a new way as its materials churn about, its crust forms and reforms, its seas and clouds pool over the rocky crust. It has enormous energy of its own and receives more energy from the Sun, which sends it light and heat. It might remind you of a chrysalis transforming a caterpillar into a butterfly, or of a chick embryo turning and growing inside its shell.

Already at this early stage the Earth begins to fit the autopoietic definition of life as it is creating its own parts, including the tiny autopoietic microbes which, as we will see in the next chapter, create the thickening atmosphere that becomes a new boundary membrane or skin. In later chapters we shall see more evidence of autopoiesis as new complex holons form within the planet's holarchy.

Had our movie shown the other planets as well, we would have seen the sharp contrast as they settled into relatively stable patterns, the solid ones dull in color, while Earth's metabolic activity brought it to life with radiant blue and green colors beneath its swirling breath of white cloud.

4

Problems for Earthlife

Imagining Gaia as a beautiful goddess dancing gives us a poetic metaphor for nature's living beauty. But real life is often hard and troublesome, as we know from our own experience. And Earth had big problems right from the time its dance of life began. In more scientific terms, we might say the probability that Gaia—our name for Earthlife as a whole—would continue to evolve was rather low during its early stages, or that a stable autopoietic Gaian system evolved only under considerable threat to its existence.

Even when its crust was already coming alive with microbes the young Earth, whirling more than twice as fast as it does now, still hissed with steam, cracked so that its lava flowed like blood and was endlessly bombarded by meteors belting in through the thin atmosphere, raising dark clouds of dust as they wounded its still tender body. The embryonic Earth's continuing life was not at all a sure thing; Gaia was not yet a secure, stable being able to maintain itself.

The constant hail of meteors, leaving craters such as we see on the Moon, was a serious threat. Though meteors may have contributed important molecules such as lipids to the formation of microbes, they might also have killed them off again. Every day these space rocks of all sizes came hurling from the sky like bullets. If nothing had happened to protect the Earth from them, it might well have ended up as lifeless and pockmarked as the Moon and our neighboring planets.

There may also have been another problem, though scientists differ on this matter. The Sun's energy was most helpful in splitting molecules so that new ones could form, but as the first microbes formed and multiplied, the strong Sunlight may have been too much for many of them to stand, putting them in need of protection from the burning part of Sunlight we call ultraviolet radiation. Some ultraviolet is good for living creatures, but too much can burn them, and our young Sun probably produced far more ultraviolet than it does today, when we are concerned about our own threat to the shield of ozone protecting us from it—a shield that did not exist at all around the early Earth, though the smoggy early atmosphere may have offered some protection.

In any case, the first microbes seem to have formed on the seafloor, in seawater or wet mud deep enough to filter out the dangerous rays. There, as we saw in the last chapter, bits of a rich soup of organic molecules and seawater were probably trapped in liposome spheres where the molecules could move about and begin new kinds of chemical cycles. These would have included, had they not already been formed elsewhere, the construction of the giant RNA and DNA molecules that became useful as a storage system for information life needed.

In the self-production and reproduction cycles that gradually evolved, RNA lined up with DNA to copy its information, then lined up with amino acids to produce the proteins coded for, which in turn helped DNA split apart to copy itself, and so on around the loop. But giant RNA and DNA molecules could be broken by ultraviolet light, and so one of life's earliest inventions, not long after reproduction itself, was the repair of DNA with special enzymes.

Early microbes were now becoming full-fledged bacteria of the type we call *archae*, simply meaning ancient. Lipid walls enclosing them permitted the entry of new raw materials and the disposal of wastes. Every living being or system has to cycle and recycle supplies. As Earth's weather cycles circulate water from sky to ground and sea and back to sky, rock is dissolved in running water and swept to the sea. Atmospheric gases are also cycled and their balance regulated. The planet's temperature is determined by all these processes, with a strong role played by its variable cloud cover.

Meanwhile, as we will see in more detail later, dissolved rock used up in forming the bodies of sea creatures ends up buried on the sea bottom in sediments pressed back into rock. Later that seafloor rock may end up as dry surface land in new plate upheavals, only to begin the cycle again.

Just so, the liposome microbes formed in the Earth's crust developed internal cycles for circulating their own supplies and carrying out the business of life. Gradually they replaced their tiny spherical capsules with larger, more flexible cell membranes and evolved into bacteria. They still depended on seawater to float supplies to them, or to float them to supplies, and to float away wastes they could no longer use. By trial and error they learned to use these supplies to grow themselves, to repair themselves when they suffered damage, and to reorganize themselves as needed, keeping records of their new discoveries in their DNA.

Every living creature must get materials and energy from its environment to form itself and to keep itself alive. What is left of these supplies after the useful parts and the energy have been taken from them, along with whatever else was part of the creature but is no longer of use to it, is waste that must be gotten rid of by returning it to the creature's environment. This is why no living creature can ever be entirely independent—it is always a holon within larger holons, including ecosystems, depending on them for its very life.

As author/scientist/philosopher Arthur Koestler put it, a holon has at once the autonomy—in Greek, self-rule—of a whole in its own right and the dependence of a part embedded within larger holons. Koestler grappled with this concept of dependence along with relative independence, referring to it as an integrative tendency, or even as self-transcendence. Let us call it a holon's *holonomy*—the rule of the greater whole or holon that must be balanced with its self-ruling autonomy. Physicist David Bohm used the word holonomy in exactly this sense when describing how the autonomy of every subatomic particle is stabilized and tempered by the rule of all other particles around it—by its holonomy. Recall our earlier discussion of bootstrap theory in physics, which also expressed this concept.

Any holon containing smaller holons, such as an Earth full of bacteria or a body made of cells, tempers the individual autonomy of its components with its *own* autonomy, which is *their holonomy*. Any individual human, for example, must transcend simple self-rule and integrate him- or herself with the rules of family and society, while human society must transcend its autonomy and integrate itself with the holonomy imposed by the autonomy of the planet. The balance between any holon's autonomy and holonomy must be worked out as mutual consistency if the holon is to survive as part of a holarchy, and it cannot survive in any other way if we accept the fundamental notion of mutual consistency as described in Chapter 2 and as illustrated in later chapters.

These concepts of embeddedness or holarchy, and of the autonomy at every level of holarchy always tempered by holonomy are extremely important to understanding how life works. We humans, for example, fight about whether to seek individual interest *or* community interest, whether to develop locally *or* globally. This is because we fail to understand life's fundamentally holarchic nature—always a dialogue among relatively autonomous embedded holons, all of which are critical to the function of the holarchy.

Negotiated Self-Interest at all Levels
TOWARD A COOPERATIVE WORLD

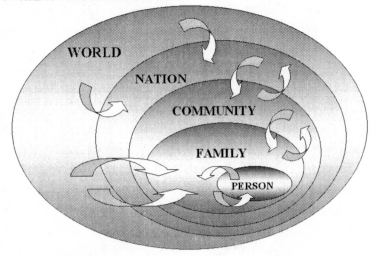

Bacteria are holons within larger holons consisting of their complex communities and even worldwide networks, as well as within their broader *ecosystems*. While we are talking definitions, let us use the term ecosystem to refer to systems of related organisms in their habitats.

Bacteria are technically called monera—the first kingdom of living things in our present evolutionary classification scheme. Monera include the archae and their later descendants of many types. (Later we will see that bacteria are also called prokaryotes, but let that come in time.) Each moneron is a single cell, and yet it is also a whole organism or creature. The tiny monera that were Earth's first creatures were thus the first relatively independent holons within the Earth holon—in Lewis Thomas' view, tiny cells within a huge cell.

Fortunately for these early monera, the sea was full of supply molecules, ranging from small dissolved rock salts to the larger sugar and acid molecules needed to build DNA and protein. So the bacteria could grow and divide and grow again, spreading themselves thickly throughout the seas. As they multiplied, winds and water driven by the Sun's energy swirled this rich chemical soup about, stirring it into ever-greater activity. So prolific were these microbes, that their colonies, including the habitats they assembled, formed entire continental shelves long before corals evolved. Even today, bacteria, or monera, are by far the most numerous creatures of the Earth.

The more bacteria there were to suck up supplies and blow out their wastes, the more the whole chemistry of the Earth changed—sometimes the worse for life, sometimes the better, as we will see.

O   O   O

Many early monera were getting their energy by breaking up supply molecules in a process we call *fermentation*. The bacteria we use to make cheese, yogurt, and wine still work the same way today. Yeasts, such as those we use to make bread, do it, too. Fermenting bacteria can be thought of as *bubblers*, since they make bubbles of waste gases, like the bubbles you see in risen bread and in cheese. Whenever you see bubbles rising in mud or stagnant waters, fermenting bacteria are probably at work.

Breaking up molecules by fermentation or in other ways frees the energy that held them together. The bubblers stored this energy in a special kind of molecule we call *adenosine triphosphate* (ATP). At first they may have found ready-made ATP molecules in their surroundings, but eventually they learned how to make them. The bubblers kept the energy-loaded ATP handy until the energy was needed for building, repair, and other work. Every living thing on Earth since then has been using the ATP energy storage system invented by the bubblers, though bacteria later discovered faster, better ways of making ATP than by fermentation. ATP is thus often called the energy currency of life.

In addition to energy, of course, the bubblers needed building supplies, and for a long time, as we said, large sugar and acid molecules were plentiful in the environment, ready to be split up or used as they were. To reproduce, some monera copied their DNA and then split themselves down the middle in the process we call *mitosis*, building two offspring monera from their own split halves. Others budded off smaller bits of themselves containing copied DNA to start their offspring. When supplies got low here and there, some bacteria learned to pack their DNA and a bit of protein into solid little spores with tough shells. These spores floated about doing nothing at all till they came to places where supplies were plentiful and they could grow into proper monera.

Over time, monera built new kinds of protein and new enzymes and invented new chemical processes and cycles, new parts for themselves, new lifestyles. More than three billion years ago, then, bubbler monera were multiplying and dividing into different strains, forming a thick soup or surface scum, living off ready-made supplies of large sugar and acid molecules. Some strains of bacteria learned to use the acid and alcohol wastes of others, and to set up efficient cycles of using one another's wastes as supplies. Some learned to make the nitrogen of the atmosphere usable by combining it with other elements. Had they not, life would have died out from nitrogen starvation, as nitrogen is one of the six basic elements needed to build living things.

Still, as competition for large-molecule food supplies increased, a new crisis developed. As if Gaia didn't have enough problems already, it began to look as though her first tiny creatures might die for lack of supplies.

But they didn't. Life is far too inventive to give up so easily.

What happened to the monera back then is rather like what is happening to us humans today. We have been making much of the energy we need to live in our human societies from the coal and oil supplies found ready-made in our environment. Now these supplies are running out, and we must find new ways to produce energy. A very important way of doing so involves the use of Sunlight, or solar energy.

This is exactly what some monera began doing as their supplies ran low. Some elements they had to have in order to build their living bodies were all around them, but like the atmospheric nitrogen, they were not in usable form. Others were hard to get at, such as the nitrogen locked into the salty nitrates of the sea or the carbon locked up in the carbon dioxide gas of the atmosphere. There was plenty of carbon and nitrogen all around, but the bubblers had to invent special ways to unlock the carbon and nitrogen and then 'fix' them by turning them into usable bodybuilding molecules.

Perhaps the bubblers' most important discovery was finding ways to harness solar energy—to trap Sunlight and turn it into ATP energy, which they did by using certain light-sensitive chemicals such as the *porphyrins* that make our blood red and the *chlorophyll* that makes grass and leaves green. They could then use this energy to split molecules of carbon dioxide gas, water, and rock salts into atoms, which could be rebuilt into food sugars, DNA parts, and more ATP for the work of growing, repairing, and reproducing. This process is, of course, *photosynthesis*—in Greek 'making with light'—the use of light in the manufacture of food.

Some of the photosynthesizing monera are called blue-green bacteria because of the color their photosynthesizing chemicals gave them. Let's call them *bluegreens* for short. Their new way of life was very successful, so they multiplied quickly. After all, the blue- greens, unlike the bubblers, needed no special supplies. Water full of dissolved rock salts was what they lived in, and the atmosphere was full of light and carbon dioxide.

O   O   O

There was only one problem: the bluegreens' wonderful new way of making their own food and energy was also creating pollution.

Both the bubblers and the bluegreens made waste gases as they worked, but light-making food from water and carbon dioxide gas produced a very poisonous waste—so poisonous that it killed living things. This poisonous waste gas was *oxygen*!

We are used to thinking of oxygen as good and necessary, as a life-giving and life-saving gas that we breathe. But for the first living creatures, it was deadly. It is oxygen that turns metals to rust and makes fires burn. Oxygen destroys the giant molecules of living things, burning them up just as ultraviolet and other kinds of radiation do. In fact, oxygen is more destructive than ultraviolet, for the large molecules needed to build the first living things could never have formed if the atmosphere had been as rich in oxygen then as it is now. So, when the bluegreens began making oxygen, they began making trouble.

Every molecule of carbon dioxide, or CO_2, is made of one carbon atom and two oxygen atoms—di meaning two. And every molecule of water, H_2O, is made of two atoms of hydrogen and one of oxygen. It takes six molecules of carbon dioxide and six molecules of water to make one molecule of food sugar. But when the sugar molecule is built from carbon, hydrogen, and oxygen atoms, it only needs twelve oxygen molecules, so six are left over as waste.

This is the oxygen that began polluting the early Earth after photosynthesis began. At first the free oxygen combined harmlessly with dissolved rock minerals such as iron, making them rust, and built itself into rock. When these crustal materials had absorbed all they could, the oxygen began piling up in the atmosphere.

It was as if a giant pump had been turned on. Bacteria were pumping carbon dioxide out of the atmosphere, using the carbon and pumping some of the pure oxygen back into it. They also pumped nitrogen out of nitrate sea salts, fixed some of it for their use, and pumped useless nitrogen gas into the atmosphere. The living Earth was bringing its own special nitrogen- and oxygen-rich atmosphere into being.

Our nearest planet neighbors, Venus and Mars, have atmospheres made almost entirely of carbon dioxide, just as was Earth's, very likely, when this great pump got going. But now our atmosphere is almost all nitrogen and oxygen—because life made it so. But how did life survive the poisonous oxygen?

Much of it *didn't* survive. Some kinds of bubblers that didn't need to be near light dug themselves down into mud where the poisonous oxygen could not get at them. Their fermenter descendants still live today by hiding from oxygen in mud or in other safe places such as the stomachs of cows, where they help digest hay, or the bead-like root nodules of peas and beans, where they fix nitrogen to enrich the soil. But many, if not most, kinds of early bacteria must have been killed as the oxygen piled up around them.

O   O   O

What was Gaia to do? Her dance of life had produced a rich array of living bacteria despite the dangers of meteors and ultraviolet light. Now most of the early kinds were dying just because some had discovered a new and better way to live. It was a lesson Gaia learned more than once, that new experimental forms of life may seriously endanger the whole dance and that other improvisations may be required to rebalance it. Gaia had no human brain to assess her experiment and think up strategies. Nevertheless, our evolving planet developed, as it still does, the kind of 'body wisdom' physiologists attribute to our bodies.

By this body wisdom, living systems operate and maintain themselves, somehow knowing what to do on a momentary and daily basis as well as in most cases when things go wrong. We are used to it in our bodies and we count on it, but to learn that the Earth behaves the same way is still news to many people. Not at all long ago it was scientific heresy to attribute intelligence to nature. Now, as we said, our worldview is evolving rapidly. Many scientists, including Lynn Margulis, the leading researcher on microbial evolution and the author of the greatest change in our 'tree of evolution' since we first devised it, see consciousness and intelligence present in the earliest microbes. If we recall that some physicists now see cosmic consciousness as the source of all matter, then it is not so surprising to find intelligent body wisdom evolving in our Gaian Earth.

Though clever species of bubblers survived by hiding from oxygen, they were no longer the main kind of monera. Bluegreens invented enzymes, which made the oxygen they produced harmless to themselves. Some also learned to make ultraviolet Sunscreens, as we make Sunglasses and chemicals to protect ourselves from Sunburn. These were able to live successfully in stronger Sunlight, where they could make plenty of food.

Others solved the problem of ultraviolet burn by living together in thick colonies. Those on top were burned to death, but the dead cells made good filters, absorbing the burning rays while letting the rest of the light reach those that needed it below. This was another way in which some lives were given for others, and a good reason for bacteria to live as cooperative life teams rather than as independent individuals.

You can see such colonies of bacteria beginning as a greenish brown scum on damp walls or muddy ground. Near the sea, they trap sand and other particles, forming thick muddy masses in shallow waters as live bacteria multiply and keep climbing toward the top. In some places we can still see this mass harden into rocks called *stromatolites*. In ancient stromatolites, the bacteria that have turned to rock can still be seen and identified.

The number of such rock colonies formed billions of years ago, sometimes extending into entire continental shelves, tells us just how successful the oxygen makers have been. This also shows clearly how rocks that rearranged themselves into living creatures can rearrange themselves back into rock.

For about two billion years—almost half of Earth's life until now—the bluegreen oxygen makers were her most successful creatures. They multiplied into thousands of different kinds all through her waters and muds, making more and more oxygen in their ever-growing colonies. And then they made yet another dramatic discovery. Some of them learned how to use the waste oxygen they created in making food molecules—they used it to burn those very food molecules for energy.

This process of burning food with oxygen is what we call *respiration*—the third way of making ATP, after the fermentation of the bubblers and the photosynthesis of the bluegreens. It is the most efficient way of all. In respiration, the destructive energy of oxygen is used to break up food molecules and thereby free both their parts and their energy for use. It is a much more powerful way to do this than fermentation. Soon other kinds of bacteria learned this method of using poisonous oxygen to good advantage. Since we call the intake of oxygen for breaking up or burning food molecules breathing, let's call the new respiring bacteria *breathers*.

Like fermentation and photosynthesis, respiration produces waste gas. But this time the waste gas is carbon dioxide—the very gas needed for photosynthesis. What an incredible new opportunity. Respiration completed a cycle by leaving a supply of carbon dioxide with which to start photosynthesis anew.

Looking closely into the green and brown living 'scum' on the tops of stromatolites, or in your kitchen sink, for that matter, with our newest and least intrusive microscopes, we are astounded! For up close, the scum, often containing vast numbers of bubblers, bluegreens and breathers, resolves itself into the most amazing cityscapes populated by all sorts of bacteria doing different tasks cooperatively while living different lifestyles. These cities look like Manhattan's skyscrapers, as one scientist put it, or like Hollywood sets for cities of the future, with buildings like tall balls on stems or cone-shaped and linked by endless canals, bridges, and other transport systems. Thus our ancient bacterial forbears built infrastructures for their communities much as we do today.

O   O   O

Let's look further at how tiny bacteria managed to accomplish so many innovations and lifestyles. Like our own human world, the developing world of bubblers, bluegreens, and breathers in the Great Bacterial Age made its progress through many technological inventions. Simple as bacteria are in comparison with later evolved creatures, they are remarkably ingenious, and we still have a great deal to learn from them. Several of our own greatest recent technological advances, such as the DNA recombination we call genetic engineering, were learned from them. Bacteria discovered this process, actually the secret of their wild success, billions of years ago, while we have just caught on and learned to get them to do it for us in ways that *we* intend.

Scientific research has shown for over half a century, beginning with Barbara McClintock's work on corm plants, that DNA reorganizes intelligently in response to specific problems faced by living organisms. It happens in life forms from microbes to very large multicelled creatures. But this evidence had to pile up very convincingly over so much time, because it so flew in the face of the official dogma that it could not be so. Evolution was only supposed to proceed by accident and 'selection.' Scientists firmly believed that changes in the forms of living creatures could happen only as a result of accidental mistakes in copying DNA, or by accidental breakage and recombination of DNA at certain times, such as when it is struck by fast-flying nuclear particles that zoom through the atmosphere and through us without our notice. But now scientists see that many, if not most, DNA changes are anything but accidental.

Modern bacteria have obviously been able to change very quickly in ways that protect them against our lethal antibiotics—in Greek, 'against life.' To do this, they have to make changes in their DNA. Life, it seems, does not just wait around for lucky accidents to solve problems and improve things, but is quite inventive, especially under survival pressure. But just how do bacteria *do* it?

We can easily see with modern microscopes that bacterial DNA is a very long complex molecule formed into a loose loop inside the tiny creature. We can also see that bacteria come very close to one another and then dissolve parts of their cell walls long enough to create a hole through which they exchange bits of DNA. One or both of them leaves this encounter with a new combination of DNA from the two though no reproduction had taken place.

This information exchange, or communication system, of ancient (and modern) bacteria is at least as remarkable as any of their other inventions and no doubt is what made the rest of their innovations possible. We are just beginning to learn how it works and to recognize it as original sex!—something we thought had been invented much later in evolution.

Sex is by definition the production of creatures by a combination of DNA from more than one individual. Every time bacteria receive bits of DNA called genes from others, they are engaging in sex by making themselves the product of two bacterial sources even though they are not reproducing. This sexual communication system apparently belongs to virtually all bacteria of all strains, so that bacteria can—and do—trade their DNA genes with one another all over the Earth to this day!

Thus these tiny ancient beings actually created the first WorldWideWeb of information exchange, trading genes as we trade our own messages from computer to computer around the world. We have speeded up their web by carrying them around the world on our ships and airplanes, to make contact in far places they might not have reached by wind and waters so quickly.

All bacteria can be thought of as one great holon with a common pool of DNA genes—a single live network or system covering our entire planet, even extending deep under its polar ice covers and into its below-surface fissures. Throughout this system the bacteria trade and recombine genes according to need and experiment. And their 'Internet' probably includes larger creatures, including ourselves, as we can see bacteria (and viruses, which may be their survival devices) coming into plants and animals to trade bits of DNA. Even before we made this discovery, we knew that no other form of life could survive today without bacteria. Why this is so will become clear as we watch the dance of life develop.

O   O   O

The young Earth's bacterial gene pool or web made it possible to spread resistance to oxygen by sharing blueprints for various protective devices, as well as to spread the use of oxygen for breathing, or burning food molecules. Of all Gaia's creatures, the blue-green breathers that harnessed solar energy were the most independent ever to evolve, and they are still going strong today, billions of years later. They make their own food *and* burn it, using only the simplest supplies. If they drift away from light, they work in the dark. They fix their own carbon and nitrogen. Living in water, they do not even risk drying up, as do the land plants that evolved long after them to carry on the double lifestyle that the bluegreen breathers invented.

The ancient bubblers, bluegreens, and breathers had invented the only three ways that living beings all over the Earth, even today, make their ATP energy currency, for as we will see, all larger creatures are their descendants. The recycling of carbon dioxide and oxygen that began with them—the simplest and tiniest of Earth's creatures—was so successful that it has been an essential part of the Gaian life system ever since. In time the two parts of this cycle—photosynthesis and respiration—became ways of life for different kinds of one-celled creatures coming up in our story. Then, much later, plants and animals evolved to cooperate in producing carbon dioxide and oxygen for each other's use, or in recycling each other's waste, depending on how you look at it.

Again we are reminded of lessons people are learning today. First the ancient bacteria solved their energy crisis by developing solar technology, then they discovered that recycling supplies is the best way to avoid running out of them.

As the oxygen piled up in and thickened the atmosphere, it not only created new problems requiring new solutions, but was *itself* a solution to the old problem of ultraviolet burn. Destructive as oxygen was to so many kinds of microscopic monera, it actually helped form a protective blanket of air around the living Earth. Just as in our ancient myth Gaia first formed the seas in her dance of life and then created a protective atmosphere, so it was in reality.

Our atmospheric blanket of air seems very thin to us. We can just barely feel it by waving our arms around in it. But what we feel against our arms would be much harder if our arms were waving much faster. Meteors move so fast that the air is quite solid to them. And rubbing hard against something solid produces heat, as you can easily demonstrate by rubbing your hand hard over a table. Meteors rub up against air so hard that the heat, together with the oxygen, ignites them and burns them up. The more oxygen there was, the more meteors burned up, until so few of them got through the atmosphere that life was much safer on Earth.

Nor were meteors the only outside dangers oxygen protected Gaia against. The oxygen in our air is made of twin oxygen atoms dancing together as free-floating molecules. As ultraviolet rays strike these molecules, they break up the pair, leaving separated twins to join other oxygen pairs as triplet molecules. Such triplet molecules are no longer oxygen gas; they are ozone—O_3. And it is very difficult for ultraviolet rays to pass through ozone because it absorbs them. When there was plenty of oxygen, a whole layer of ozone collected in the middle of the atmosphere, shielding the Earth from dangerous amounts of ultraviolet radiation.

Microbe-produced oxygen probably even played a role in preventing the seas from drying up, because atmospheric oxygen can trap evaporating lightweight hydrogen as water, thus preventing it from escaping into space and allowing it instead to fall back into the seas in the form of rain. Methane-producing fermenters may also help hold the oceans onto Earth, as atmospheric methane decomposed by ultraviolet rays creates the tropopause lid, which is another barrier to the escape of hydrogen into space.

O   O   O

From this early history of the Gaian dance of life we can see that great problems are great challenges, and that living things are very inventive when faced with challenges. Maybe that is one of the most important things we can learn from evolution.

It remains to be seen whether we humans will prove as creative as ancient bacteria in the face of the problems *we* create. Over billions of years, most of the carbon dioxide was pumped from Earth's atmosphere by photosynthesis, while nitrogen, oxygen, and rarer gases produced by living creatures replenished it. Over this long time, life worked out exactly the right balance of gases that was best for it. Now we are changing that balance in dangerous ways.

Our use of coal and oil, for example, is creating a very serious double problem. Not only are we using up these important fuel supplies but we are also polluting the atmosphere with too much carbon dioxide in burning them. Coal and oil are made of ancient forests that built much carbon into their plant and animal life. As they were pressed underground over time, the carbon was buried and transformed into natural fuels. That is why we sometimes playfully call oil 'dinosaur blood.' When we dig up these fossil fuels and burn them, the carbon is released back into the atmosphere as carbon dioxide.

At the same time, we are also burning today's growing forests to clear land for our use. This releases even more carbon into the air while killing the very plant life that uses up carbon dioxide to make oxygen, thus preserving the balance.

Billions of years ago oxygen was the great danger. Now the danger is too much carbon dioxide. The result, among other things, is that our planet is heating up, for too much carbon dioxide prevents its normal loss of heat through the atmosphere. When our own bodies heat up in this way, we have a fever. If *we* don't solve our energy and production problems very soon in ways that are healthful for life, the Earth will have to solve the problem itself, restoring its balance as best it can. Atmospheric carbon dioxide is rapidly approaching levels it apparently reached previously just before the ice ages. Perhaps Gaia will cool her man-made fever with a new ice age, destroying most of what we have built and forcing us into retreat—like the ancient bubbler bacteria—to safer environments.

Inconvenient as another ice age would be, we at least know humans have survived a number of them by moving to the tropics where new land is exposed as ocean water is removed to form snow and ice. Far worse would be Gaia's other alternative, which is now looking ever more likely: to reset her thermostat at a higher planetary temperature, thus regaining the Gaian system's stability as a whole at our expense. Humans, other land animals, trees and other land life, would all succumb to the increased heat and the loss of almost all dry land if polar caps melted and flourishing oceans rose dramatically in a heat age.

This is what we are learning: to understand that the Gaian life system has evolved in such a way that it takes care of itself as a whole, and that we humans are only one part of it. Gaia goes on living, that is, while her various species come and go. We used to believe that we were put here to do whatever we wanted to with our planet, that we were in charge. Now we see that we are natural creatures which evolved inside a great Earthlife system. Whatever we do that is not good for life, the rest of the system will try to undo or balance in any way it can. That is why we must learn Gaia's dance and follow its rhythms and harmonies in our own lives.

5

The Dance of Life

It was in the search for life on other planets that we discovered what a live planet is and that we ourselves are part of the only live planet in our solar system.

The first astronauts to see the whole Earth with their own eyes were astonished by what they saw. Although they couldn't see any of the living creatures they knew to be on it, the Earth itself looked very much alive—like a beautiful glowing creature pulsing or breathing beneath its swirling, veil-like skin. Their pictures helped us in imagining Earth's evolution as a film.

Scientists, of course, cannot simply trust the way things look. After all, science was built on the discovery that the Earth is *not* the unmoving center of the universe, much as it looks to be just that. Nevertheless, it was seeing our planet from afar for the first time, and noting how its appearance differed from other planets, that inspired new ideas and studies of Earth, such as Gaian science.

Long before we saw our planet in this new way, scientists had adopted the view that the Earth with its various environments is a

68

nonliving geological background for life, living creatures having evolved upon it by accident and having adapted to it by natural selection. The Scottish scientist James Hutton, who is remembered as the father of geology, was virtually ignored when, in 1785, he called the Earth a living superorganism and said its proper study should be physiology. A century later the Russian philosopher Y. M. Korolenko told his nephew, Vladimir Ivanovich Vernadsky, that the Earth was a live being, and though it is not clear that Vernadsky believed this himself, his studies of Earth took a very different view of life than did those of other scientists.

Vernadsky called life "a disperse of rock," because he saw life as a chemical process transforming rock into highly active living matter and back, breaking it up, and moving it about in an endless cyclical process. Vernadsky's view is presented in this book, as we say life is rock rearranging itself—like music come alive—packaging itself as cells, speeding its chemical changes with enzymes, turning cosmic radiation into its own forms of energy, transforming itself into ever-evolving creatures and back into rock. This view of living matter as continuous with, and as a chemical transformation of, nonliving planetary matter is very different from the view of life developing on the surface of a nonliving planet and adapting to it.

While Vernadsky's view stimulated much research in the Soviet Union, it never became widely known in the West. The biologist G. E. Hutchinson was one of the very few Western scientists of this century who took an interest in and promoted Vernadsky's view that life is a geochemical process of the Earth.

Later the independent English scientist James Lovelock, who assisted NASA in its search for life on Mars, shocked the world of science by suggesting that the geological environment is not only the product and remainder of past life but also an active creation of living things. Though he did not know Vernadsky's work at the time, he said that living organisms continually renew and regulate the chemical balance of air, seas, and soil in ways that ensure their continued existence. He called this idea—that life creates and maintains precise environmental conditions favorable to its existence—the Gaia hypothesis, at the suggestion of his Cornwall neighbor, the novelist William Golding.

The Gaia hypothesis is now recognized as Gaia theory, but it is still controversial among scientists. Lovelock, like his predecessor Hutton, calls Earth-as-Gaia an organism or superorganism and claims its proper study is physiology. Yet he also calls Gaia a self-stabilizing mechanism made of coupled living and nonliving parts—organisms and physical environments—which affect one another in ways that maintain Earth's relatively constant temperature and chemical balance within limits favorable to life. Lovelock describes this mechanical system as a cybernetic device working by means of feedback among its coupled parts. Thus it maintains Earth's stable conditions in the manner of a thermostat-controlled heating system that maintains house temperature, or an automatic pilot that keeps an airplane on course. This concept of Gaia as a cybernetic device has been far more acceptable within the mechanical worldview that is still strong among scientists than is the concept of Gaia as a live organism, though this is changing.

For Lovelock, *organism* and *mechanism* are equally appropriate concepts, but in fact the two concepts contradict each other logically, and this causes confusion around the whole issue of Gaia theory. The concept of *life*, by any definition, including the autopoietic definition (of self-producing and self-renewing living systems, as introduced in Chapter 3), is not logically consistent with the concept and reality of mechanism.

For one thing, life cannot be *part* of a living being; life is the essence or process of the whole living being. If Gaia is the name given to the living Earth, then it would be as meaningless to say that life creates its own environments or conditions on Earth as it would be to say that life creates its own environments or conditions in our bodies. Life is the process of bodies, not one of their parts, and in this book we maintain that the same is true for Gaia-Earth—that life is its process, its metabolic system, its particular kind of working organization, not one of its parts.

We can of course say that organisms within Gaia create their environments and are created by them, in the same sense that we say cells create their own environments and are created by them in our bodies. In other words, there is continual and mutually creative interaction between holons and their surrounding holarchies. But, in this book's story, we do not divide living bodies or holarchies into life and non-life.

If we accept the autopoietic definition of life, we see another contradiction between Gaia as a living being and Gaia as a mechanical system in which life and non-life are coupled parts. An autopoietic system is self-producing and self-maintaining. It must constantly change or renew itself in order to stay the same. Your body renews most of its cells within each seven years of your life, for instance, and its molecules are turned over far faster. No mechanism can do this, because it does not invent and build itself, it must be invented, built and repaired by external beings. It is not autopoietic, it is allopoietic (not self-produced, but other-produced).

A mechanism cannot, and therefore does not, change itself by its own rules, and that one fact points out the essential difference between living systems and mechanical ones, including even the most sophisticated computers and cybernetic robots. All of them must be programmed by outsiders to do what they do, no matter how intelligent they appear to be. We will say more on this subject later, especially in Chapter 15. For now, let us just note the contradiction that arises if we define Gaia at once as a living organism and as a cybernetic device, because this contradiction is causing confusion about Gaia theory among scientists.

The position of this book, then, is that the Earth meets the biological definition of a living entity as a self-creating autopoietic system, and that only limited aspects of its function—never its essential self-organization—may be usefully modeled by cybernetic systems. For example, we can usefully model aspects of our own physiology (for instance, temperature regulation, or blood pumping) as cybernetic feedback systems, knowing all the while that these systems would not function like machines apart from their embedding holons' physiology.

Notice that calling the Earth alive, by definition, is more than proposing a new metaphor to replace mechanism. It is also different from proposing a Gaia hypothesis or a Gaia theory. There is nothing to be proven once we decide that Earth fits the autopoietic definition of life, as it simply revises our conceptualization from mechanism to organism. And, as such, it provides fruitful ground for many new hypotheses and theories about how its physiology works. Note, by the way, that autopoiesis as a definition of life does not include growth or reproduction, though these are features of many living entities. One can be alive without reproducing, perhaps an important recognition in an overpopulated world.

O   O   O

Let us look now at the Earth as the self-creating living planet we are calling Gaia to distinguish it from a nonliving planet with life upon it. We have already seen how magma is constantly transformed into crust, how crust is transformed into microbes and organisms, how these are turned back into crust and magma to complete the ongoing cycle of self-creation.

Lovelock's first clue to Gaia came to him when he was comparing the atmospheres of different planets. The atmospheres of the other planets in our solar system all make sense chemically as stable mixtures of gases. Only Earth has an atmosphere that is quite impossible by the laws of chemistry. Its gases should have burned each other up long ago.

If they had, Earth would have no living creatures. And of course it *does*. They make and use almost the entire mixture of gases we call the atmosphere, ever feeding it new supplies as they use it and as it burns itself up chemically. This activity of living things always keeps the atmosphere in just the right balance for the life of Earth to continue. We can compare it to the activity of our cells in producing, using, and renewing the blood, lymph, and intercellular fluids flowing around them.

Living creatures, for example, produce four billion tons of new oxygen every year to make up for use and loss. They also make huge amounts of methane, which regulates the amount of oxygen in the air at any time, and they keep the air well diluted with harmless nitrogen. In fact, the Gaian atmosphere is held at very nearly 21 percent oxygen all the time. A little more and fires would start all over our planet, even in wet grass. A little less and we, along with all other air-breathing creatures, would die.

Every molecule of air you breathe, with the exception of trace amounts of inert gases such as argon and krypton, has actually been recently produced inside the cells of other living creatures! Thus the atmosphere is almost entirely the result of the constant production of gases by organisms. If they stopped making and balancing the gases of our air, the atmosphere would burn itself up rather quickly. And if living things didn't turn salty nitrates into nitrogen and pump that nitrogen into the air, the seas would become too salty for life to go on in them, and the atmosphere would lose its balance. The right balance of chemicals and acid in the seas and in the soil, and even the balance of temperature all over the Earth—all of the conditions necessary for the life of our planet, that is—are regulated within the planet as they are in our bodies.

Our Sun has been growing larger and hotter ever since the Earth was formed, yet the Earth has kept a rather steady temperature, in much the same way that a warm-blooded animal keeps a steady temperature while things get cooler or hotter around it.

Old attempts to explain how geological mechanisms might regulate the Earth's temperature are giving way to new explanations of how a live planet does it. Part of the complicated system involves regulating 'greenhouse gases,' such as carbon dioxide and methane, which trap solar heat; another part involves controlling the amount of cloud cover to let in more or less Sunlight. Perhaps the Earth even creates ice ages to cool its fevers.

In our own bodies, there are always things going on to upset the balance of oxygen or salt or acid in our blood and cells. Yet the parts of our living body work together constantly against these upsets of balance. Just so, it seems that the parts of the Earth work together to help it recover from its own imbalances, though as yet we know little about how this is done.

Although we have learned much about the ways in which the complexly coordinated systems of our own bodies function, we

can hardly even dream of knowing everything involved in building and running such systems. We seldom reflect on the fact that
our bodies work without asking anything of our aware, thinking
minds. We need not even know consciously what is going on,
much less having to think or plan or do anything about it. And a
good thing this is, because we would most certainly mess up our
bodies' wonderful work if we interfered in it in an attempt to control it ourselves. Lewis Thomas, the popular science essayist and
physiologist mentioned earlier, has said that for all his physiological
knowledge, he would rather be put behind the controls of a jumbo
jet than be put in charge of running his liver. Any one of our
organs is more complicated by far than the most complicated computer we've invented, yet it knows how to run itself, repair itself,
and work in harmony with all other organs.

We are just beginning to understand cosmic consciousness and
are even farther behind in understanding the nature of consciousness in living cells and bodies that clearly know what is good for
them. They know just how they should be balanced as well as how
to do the balancing. This still mysterious 'body wisdom' or intelligence seems to exist throughout nature, somehow evolving the
kind of consciousness familiar to us as humans. Eastern human
cultures have studied non-physical aspects of nature for thousands
of years; western science is just beginning to do the same.

The sooner we recognize and respect Gaia as an incredibly
complex and demonstrably intelligent self-organizing living
being, the sooner we will gain enough humility to stop believing
we know how to manage her. If we stay on our present course,
clinging to our present belief in our ability to control the Earth
while knowing so little about it, our disastrously unintelligent
interference in its affairs will not kill the planet, as many people

believe, but it may very well kill us as a species, as we are already killing so many others.

O   O   O

Starting with physicists' current view of cosmic beginnings, we have seen that the universe has tremendous energy to spend, and that it spends this energy evolving itself into ever more complicated patterns, including those we recognize as alive. We have come to believe that the total useful, or working, energy of the universe—according to the laws of physics, in particular the law of entropy—is gradually running down. Yet living creatures collect, store, and increase working energy wherever they find it, thereby violating this law. To keep the laws of physics consistent, scientists believe that in increasing energy locally living beings must be decreasing the energy of their environment at an even greater rate. Only thus would they satisfy the overall demands of the entropy law, otherwise known as the second law of thermodynamics, the law, which says that things are running down as a whole. This implies that living things must use up and thereby degrade their environment, making it ever less useful to other living things.

On our planet this would mean that each form of life gradually uses up or degrades its environmental supplies until it chokes itself off and dies. Indeed it seems that some living creatures sometimes behave in just that way, as did the first bacteria when they used up the ready-made sugars and acids in their environment, and as we humans do when we use up and destroy our natural resources. But when one kind of organism creates such a crisis, the living Gaian system as a whole seems to find a solution.

What about the planet as a living whole? Does it degrade its environment as it organizes its complexity? It is very dependent on solar energy, to be sure, but the Sun does not burn up faster because the Earth uses its energy, and the waste heat given off by the Earth cannot be construed as degrading its cold space environment. Over billions of years—surely a more than adequate test for the law of entropy—our Gaian planet has continued to self-organize in ever greater complexity, recycling its supplies without running down the way mechanical systems do.

We could, of course, think of entropy as the *catabolism* side of a metabolic cycle: *anabolism* building things up and *catabolism* breaking them down. We have already seen several ways in which the Earth's metabolic cycles work. Physicists are now taking about black/white holes—destroying matter in their black aspect; creating it in their white aspect. Some even believe there may be mini black/white holes at every conceivable point in spacetime, creating and destroying matter simultaneously and continually.

Earth *has* had occasional big shake-ups of destruction in its evolution—we call them *extinctions*. In fact, the five major extinctions we can document gave rise to great bursts of new creativity though up to 90 percent or more of its species died out in them. Let us hope the sixth great extinction, which we humans are now causing, according to biologists polled by the American Museum of Natural History, will cause new creativity in ourselves, rather than our own extinction. At present it is proceeding faster even than the extinction 60 million years ago which did in the dinosaurs, through no fault of their own, since that one was caused by a huge meteorite plunging into the Earth and severely altering its climate. In any case, extinctions can hardly be construed as the working of entropy.

Certainly it would seem that the entropy law of thermodynamics, discovered to explain how nonliving 'closed' mechanical systems such as steam engines work, can tell us little about living systems. So, again we run into a contradiction between mechanics and organics. Many non-scientists, many readers of this book, probably find it strange that scientists *do* try to explain life in mechanical terms, feeling intuitively (and rightly) that there is something wrong with the whole idea. In later chapters we will see how the mechanical worldview of science and society came about and how it is now changing.

O   O   O

Recent discoveries in physics strongly suggest that the nature of the universe was from the beginning such that it would come alive however and wherever possible. Perhaps planets are to our galaxy something like seeds and eggs are to multicelled Earth creatures, in that far more of them are produced than can actually form new living beings. And perhaps, like the cells in our own bodies, the 'cells' of the universe, in the form of star systems or planets, may be alive for a time and then die. After death, their components may be recycled—in other words, the energy locked up in their atoms and molecules may be used again by some other part coming alive and needing supplies to develop. Those parts of the universe that seem most lifeless to us may be something like its skeleton—providing a framework as does the core of the Earth in supporting its living surface, or the deadwood forming most of a redwood tree under *its* living surface.

If we agree that nature is not mechanical but organic, why should we not understand the energetic motion of the very first whirling shapes in the early universe as the first stirrings of that self-organizing process leading to living organisms? The spiraling pre-galactic clouds, composed of spiraling atoms, held themselves together, drew in more matter-energy from their surroundings, built it into themselves, and lost energy again to their surroundings. In this process of energy exchange they evolved into new, more compli-cated forms. By the time we get to galaxies and to fully formed stars within them, the dance toward life has become quite complicated already. We are only beginning to discover how complicated are the structure and process of our own Sun star, and we still have much to learn about the way our own planet rearranges its matter into those lively chemical patterns we all agree to call living organisms.

Earth, it now appears—though we still search—is the only planet or moon in our solar system that had just the right size, den-sity, composition, fluidity of elements, and just the right distancing and balancing of energy with its Sun star and satellite Moon to come alive and stay so. Yet its life is a result of this fortunate con-fluence of conditions, just as the development of a plant or animal embryo is. Our living Earth is likely no more a freak accident than is the seedling that grows or the frog egg that matures. All are the inevitable result of right compositions and conditions.

Some scientists believe the conditions of Earth were *so* special that Earth is a rare phenomenon, perhaps the only such planet in the universe. But there is no better reason to believe this than there is to believe that living planets are as common in the universe as are the successful seedlings and hatchlings of Earth. And if this is so, there are billions, maybe trillions, of other live planets in the bil-lions of galaxies, each with their billions of star systems. Surely we are *not* alone.

To continue looking at Earth as alive, we can note that the only part of the Earth more energetic than living creatures is the lava erupting or oozing through its crust. Yet, most of that energy is quickly lost as heat pouring into the atmosphere, while living things recycle their energy within and among themselves and from one generation to another. The living matter of the Earth contained in all its creatures, according to Vernadsky's measurements, is up to a thousand times more active, more energetic, than the rocky crust from which they evolved. Hardly an example of the decreasing energy predicted by the entropy law. Where did all this energy come from?

The giant molecules from which the first creatures formed themselves were produced by powerful solar and lightning energy or from Earth's hot core, and some of that energy got locked up within them. The creatures formed from these molecules released this energy in various ways, often creating new energy in doing so. They also learned to use solar energy directly as we have seen, maintaining themselves and producing an oxygen-rich atmosphere in the process. Oxygen-burning respirers get their energy by consuming fermenters, photosynthesizers and one another. Organisms can thus convert stored energy or direct solar energy into other useful forms of energy—the energy of motion, of heat, of chemical reaction, even of electricity. Meanwhile, the atmosphere they created regulates the kinds and amounts of solar radiation available, keeping it within appropriate bounds. In Lewis Thomas' words, Earth seems to be a creature "marvelously skilled in handling the Sun."

Meanwhile, the raw materials of the Earth's interior spew or well up as new rock to be transformed into living matter, while old living matter, dead and compressed back into sedimentary rock, sinks back into the soft mantle at the edges of tectonic plates. On the Earth's surface, scientists have a hard time finding any rock that has not been part of living organisms, that was not transformed into living matter before it became rock again. We will see examples of this process later.

Thus the molecules in virtually all of the atmosphere, all of the soils and seas, all of the surface rocks and much of the underlying, recycling magma, have been through at *least* one phase in which they were within living creatures! It is easier to distinguish between life and death than between the domains of life and non-life we have assigned to biologists and geologists, respectively. In fact, virtually every geological part or feature of Earth we can find is a product of our planet's life activity. Further, living organisms have invented 99.9 percent of all the kinds of molecules we know, almost all of them back when bacteria were the only creatures around, a few billion years ago.

O   O   O

What confused us for so long—kept us from seeing that our planet is alive as a whole—is at least in part our own human space (size) and time perspective. We easily see ourselves, and many kinds of plants and animals, as wholes—separate from one another and from their surround. We have had as hard a time recognizing ourselves or them as parts of a single being as we had recognizing that we ourselves are made of separate cells. In one instance we saw the parts more easily than the whole; in the other we saw the whole more easily than the parts.

If we had a magnifying glass powerful enough to let us see everything in the world around us at the level of molecules, we would see life in the energetic molecular dance of chemical reactions and recombinations—the dance that weaves molecules into new patterns, some livelier than others. Instead, our experience comes through eyes that see life as a collection of separate plants and animals. This makes it hard for us to see them as parts of their environment, much less as parts of a whole living planet. Yet when we see the whole Earth from far enough away to show it on a movie screen and speed up its rotations, it *does* look alive, though we can no longer see its separate plant and animal parts. We have no way of seeing our world of life-within-life at all its size levels at once, but we can use our minds to put information about different levels together and understand its living holarchy of holons.

The smaller living holons or organisms within Gaia grow and reproduce, so we have come to think of growth and reproduction as essential features of living beings. The autopoietic definition of life, remember, does not include them as essential or defining features; rather, they are consequences of the autopoietic life process—something that may or may not happen, as when people do or do not reproduce. Therefore, the argument that the Earth cannot be alive because it does not grow or reproduce does not hold.

Cells, as we saw, are the packages in which living matter contained itself as it individuated. Cells contain and connect autopoietic systems by enclosing them in open boundaries or membranes of their own making that allow materials and energy to be exchanged with the environment, as does the self-produced atmospheric membrane of the Earth. In a sense, the whole Earth is a giant cell within whose boundary membrane other smaller cells evolve, multiply, die, and are recycled in such a way that the whole need not grow. This is a wonderfully efficient way to make living planets possible when they have only occasional meteor or comet impact for material nourishment.

Because our perception has been so focused on separate organisms in their physical or social environments, we tend to see insect animal, and human societies, as well as whole ecosystems, as collections of individuals that have come to live and function together. It is actually more appropriate to see that such collections have always functioned as wholes, never separated into completely individual and independent beings. Some are relatively more or less independent than others, but all their complex forms and ways have evolved within a single system, just as our cells evolved their separate functions within an inseparable whole. Their connections with their species fellows and with their ecosystems are always as holons within holarchies, up to the whole Gaian planet. These interconnections were never broken and cannot be, just as our cells cannot break their connections with their organs or their/our whole bodies.

Some scientists trying to understand Gaia as a collection of separate organisms mechanically coupled to their nonliving environments, get bogged down in arguments. How could organisms—collectively called the *biota*, or life—actually have joined forces on purpose to control the conditions of their *abiotic*, or nonliving, environment in their own interests? How could all bacteria—assuming the Gaian mechanism was formed when there were no other creatures—get together, they ask, to work cooperatively and purposively for their own good?

Official science does not recognize purpose as part of nature. It *does* view nature as mechanical, and mechanisms are, by definition, the purposeful inventions of their inventors. But historically, science threw God-the-inventor-of-nature out, we recall, while retaining the idea of nature as mechanism (more on this in Chapter 15). Scientists thus argue a logical contradiction: that nature is mechanical but has no creator and no purpose. Its machines are taken to have assembled themselves by accident. If this worldview seems hard to accept, rest assured it is now changing.

When we see living entities as self-ruled autopoietic systems evolving without an external creator God, we see that they simply evolve wherever they are not prevented from so doing, wherever their energetic development is mutually consistent with whatever else is going on around them. This scientific view is perfectly consistent with seeing nature as conscious and intelligent, in fact, suggests that strongly. Alternatively, we can choose to identify all of self-creating nature with the concept of Creator or God, thus ending the split between science and religion.

No one argues about whether or not our bodies regulate our temperature on purpose—we simply accept that they do so because they evolved that way. In the same way, bacteria did not have to assemble for the purpose of controlling their environments. We have seen that all bacteria are living matter transformed from Earth's rocky crust and packaged in open boundaries that keep them functioning as a single system. They are not separate from one another or from the crust; they are not parts of an assembled mechanism but part of a single Gaian life process we can call *geobiological*. Earthlife has evolved to do what it needs to do in order to preserve itself as naturally as we do and with no more or less purpose than we find in our own bodies.

If Gaia is a single live planet, why did its rock rearrange itself into such an astounding variety of individual creatures? Why not just a planet holon, instead of a planetary holarchy of holons?

We might as well ask why the first gas clouds sorted themselves into individual galaxies, and the galaxies into stars and planets and other space bodies. One answer, as we now begin to understand, is that life becomes ever more stable as it becomes more complex. Mechanical systems may be more vulnerable to breakdown as they become more complex, but the opposite seems to be true of living systems.

The Gaian division of function among different species is like the division of labor among the types of cells and organs in our bodies, which function efficiently through their combined work. No place on Earth—from the barest mountaintop to the deepest part of the sea—has fewer than a thousand different life forms, mostly microbial, doing different things to keep it alive and evolving. If a planet does come alive, it would seem that it must come alive everywhere, not just in patches.

Scientists are only now beginning to work out the physiology of our Gaian planet—to understand why the introduction of a single new species into a complex environment can make that environment ill, just as the introduction of a single species of disease microbe into our bodies can make us ill. They are only now coming to understand why the destruction of an environment such as tropical forest can unbalance the whole planet, just as removing an organ from our bodies can unbalance us. Yet we are also discovering that Gaia's incredible complexity makes her even tougher and more resourceful than we are. We are far more likely to choke our own species off by destroying our environment than we are to kill Gaia. Gaia's evolving dance of life will continue with or without us.

The word *evolution*, when used in the field of dance, means the changing patterns of steps, the transformative movements, in any particular dance. A dance thus evolves as its movement patterns, and perhaps its costumes, change into new ones. A good dance has overall harmony, each of its movements contributing to the entire piece. In exactly this sense, the evolution of Gaia's dance—of Earthlife— is the changing patterns of steps, ever transforming the interwoven self-organization of all creatures and their habitats over time. This is a very different view from that of biological evolution as survival of the fittest.

We see that Gaia's dance is endlessly innovative. Trying out new step patterns in a dance is called improvising. Improvisational dance is not planned out in advance. Rather the dancers improvise as they go, testing each new step or configuration for its fit with other steps and with the whole dance pattern. Gaia's dance seems to have evolved by such improvisation, the working out of basic steps used over and over in ever-new combinations. Is cosmic consciousness expressing its creativity in all possible ways as it seeks harmonious patterns?

In Gaia's dance, all creatures, from the first bacteria to trees and ourselves, have built themselves from DNA and protein molecules. The very complex patterns of these giant molecules are almost entirely made of only six kinds of atoms, as we saw: hydrogen, carbon, nitrogen, oxygen, phosphorus, and sulfur. And as we also saw, many kinds of atoms other than hydrogen were created all over the universe inside stars. As particles had combined to form atoms, the originally abundant hydrogen atoms had fused into heavier element atoms, and elements had formed themselves into molecules, which formed themselves into creatures.

We saw that there are very few kinds of protein or other molecules on Earth today whose patterns the ancient bacteria had not already invented billions of years ago. Nor have any new basic life processes been developed since bubblers, bluegreens, and breathers invented theirs. In other words, evolution since then has been a matter of rearranging not only the same atoms but also the same molecules and life processes into an endless variety of new creature patterns. This, then, is Gaia's dance—the endless improvisation and elaboration of the same elegantly simple steps into the ever-changing awesomely beautiful and complex being of which we are the newest feature.

6

A Great Leap

Little more than one century ago, priests told people that the Earth was a few thousand years old; a few decades ago, scientists believed that life on Earth began only slightly more than half a billion years ago. Now we know that the Earth's skin was already swarming with fully evolved monera well over three and a half billion years ago. Two billion years, from about 3.7 to 1.7 billion years ago—almost half the Earth's life—belonged solely to bacteria. This chapter is the story of the dramatic leap into the other four kingdoms of life, a leap that took place about 1.7 billion years ago and which holds important lessons for humanity today.

Before we go into that story, however, let us recall the magnificent work of the bacteria in preparing the way for their own evolution into other forms of life. Let us recall that they invented all life's ways of making a living and created the conditions for life that we huge latecomers enjoy today. Without the bacteria, the Earth's atmosphere would be unbreatheable and its crust would have remained a cratered desert of glassy rocks. Without the activity of bacteria, even the oceans, as we saw, would have gassed off our planet.

When we left the story of evolution, the bluegreens were turning solar energy to use in making food, and turning food energy to use in the work of life. Their waste gas, oxygen, together with other waste gases, such as the methane made by the older bubbler bacteria, piled up, creating a new kind of atmosphere. Gaia had solved some big problems and was thriving.

The bluegreens tried out all sorts of new shapes and configurations for their one-celled bodies—tiny balls, big blobs, long strings of individuals joined end to end, and even great sheets of them all stuck together with jellylike stuff and looking like seaweed. Some lived in large colonies, branching out into shapes that plants adopted when they evolved much later. Some even made spores in ways that remind us of plants making seeds. Perhaps the coded plans for such shapes and parts were stored in bacterial DNA and passed on for many millions of years until true plants evolved and found use for them. Other kinds of bacteria lived on their own as threadlike whips, lashing themselves from place to place. This was another early-evolved structure that prefigured a way of moving that would evolve later in the dance with flexible backbones and muscles.

Besides trying out new shapes and movements, the monera evolved a division of labor that streamlined individual bacteria by reducing the amount of DNA and equipment each had to carry. Various kinds of monera became specialists at particular jobs, such as respiration, photosynthesis, or fixing nitrogen gas, yet all of them had access to the whole bacterial gene pool because they never lost their ability to trade DNA in their WorldWideWeb when necessary or desirable.

The new, streamlined specialists spread out into new habitats. Some did well in freezing cold waters; others lived in very hot springs. Some blew about in the air they had helped to create, then settled far from where they began. Some even found it possible to live and multiply on land, eating their way into rocks, where they started the processes that would eventually turn them into soil. But the more specialized the monera were, of course, the less independent they were, the more they depended on one another. Oxygen users now needed oxygen makers; food eaters needed food makers, nitrogen builders needed nitrogen fixers, and so on. In specializing, the monera were evolving what we call food chains, or ecological systems, in which each species provided for and took from others.

It seems that nature must always work out a balance between the independence and *inter*dependence of individual creatures— between their autonomy and their holonomy. Specialization— whether in human society, ecosystems, multicelled creatures, or bacterial networks—is a feature of whole systems that makes them more versatile and efficient through the interdependence it creates among parts. Specialization brings variety into the life dance, but increases holonomy at the expense of autonomy, since it increases interdependence. This balancing of autonomy and holonomy is very important to understand if we are to learn to manage our human affairs as well as Gaia has worked out hers.

We saw the bacteria multiplying to ever greater numbers, discovering ever more and different ways of surviving and making their living, adapting themselves to geological changes in their environment as well as to the changes they themselves brought about. Before the ozone layer had gathered, strong ultraviolet light often damaged their DNA, but they invented splicing enzymes to repair the damage, and they learned to share this information, as well as other information stored in DNA plans, with each other. Thus the genetic plans for many variations on their organization were kept available, to be borrowed and copied from one another.

The bacterial population as a whole could therefore respond to emergencies such as chemical changes in the environment by drawing on and quickly spreading the genes that best helped them cope with those emergencies. Nowadays, part of their coping in this manner is their developing resistance to man-made antibiotics, or learning to digest new foods in changing environments. We also observe, in the frequent failure of our genetic engineering, that organisms recognize and remove—edit out—implanted genes.

Huge teams of specialist bacteria were dividing up tasks, recycling Earth's materials, learning to balance the whole dance of life. But as specialists they also ran into trouble now and then. We can imagine that blue-green food makers needing light, for instance, must have found themselves stuck too long sometimes in dark places, and bubbling or breathing food eaters must have found themselves short of food because there was so much competition for it, so many mouths to feed, so to speak.

Ultraviolet light reaching the Earth had helped produce the sugars and other food molecules on which bubblers and breathers depended. As the ozone layer grew and began to screen out ultraviolet rays, this production was severely limited, increasing the competition for food to the point of crisis. The challenge of worldwide hunger seems to have pushed the monera to rediscover some old steps and patterns in their dance of life and weave them together into a new pattern that produced a very great leap in the dance of evolution.

This leap was accomplished when the ever more specialized bacteria got together within the same walls, where they could use their various ways of making a living cooperatively. In doing so, they evolved a very large and sophisticated new kind of cell—a kind of cell so different from the bacterial moneron that it is more closely related to us than to any of its own ancestors.

O   O   O

These new cells—on the average a thousand times bigger than bacterial cells—formed a second kingdom of life to join the monera: a kingdom of life we call *protista*. The name comes from a longer word, *protoctists*, Greek for first builders. Protists, like monera, are single-celled creatures, though much larger. Later in our story, they will go on to build multicelled creatures, but for now let us look at them as multi-creatured cells.

Although the big new protists, once evolved, were smoothly run cooperative ventures, they did not start out that way. In fact, the new step in evolution almost certainly began quite uncooperatively in the desperate search for food. With the growth of bacterial populations and a developing ozone layer, the time when free food lay or floated all around came to an end. Natural death had not yet been invented to recycle materials, as bacteria do not necessarily or normally die and dissolve into reusable parts. Rather, parent bacteria split themselves to become their own offspring, and ever more offspring made ever greater demands on the food they needed in order to build themselves to full size.

Earth's crust had come alive by packaging ever more of its atoms and molecules into bacteria, many of which depended on ready-made food supplies that were now limited by the new atmosphere. The planetary process of coming alive was thus in danger of choking itself off by overcrowding and lack of food. It is not impossible that something like this happened to Mars—that Mars once came alive and then died for lack of supplies, or lack of a means to recycle supplies efficiently enough to keep its early creatures healthy.

On Earth, the evolution of giant cell cooperatives probably began when tiny energetic breather bacteria began forcing their way through the walls of larger bubblers to get at their rich molecules—not entirely unlike the way in which we humans invaded each other's kingdoms and countries to get at supplies and raw materials.

The problem with this approach was—as it still is—that eventually the invaders run out of supplies again, having eaten up their hosts. In the long run, invaders have little more to gain than their victims. But this crisis was a new challenge to life, and life proved as inventive in this situation as it had been in the face of previous problems.

Unable to get rid of, say, the invading breathers multiplying within them, the big bubblers seem to have negotiated an agreement with them that was of benefit to both parties. Perhaps in return for feeding on the bubblers' molecules, the breathers gave the bubblers some of the ATP energy they could make so much faster. This is not unlike the deals made between countries, when, for example, a rich country offers electrification in return for a Third World country's food products. As we have learned, it is not easy to make such deals truly beneficial to both sides. Such a bacterial agreement would have helped the big bubblers repair themselves or make extra molecules to feed the breathers in return for supplying them with plenty of energy.

Arrangements of this kind must have been worked out successfully, because the first protist fossils from about a billion and a half years ago show lots of breathers inside single big cells that were apparently healthy. This tells us that the breathers multiplied to large numbers within the host cell without damaging it. When such swelled cells divided, half the breathers went to one side, half to the other. And so the cooperative exchange of food for energy continued in ever more descendant cells.

Later such cells seem to have taken in other partners, such as bluegreen specialists in photosynthesis. The big hungry bubblers probably ate the bluegreens, but they apparently resisted being digested. Eventually another cooperative agreement was reached: the bluegreens would make food molecules, which the bubblers and breathers needed, by photosynthesis. This cooperative arrangement allowed the giant cells to make ATP energy in all three ways within the same cell wall and thus to survive through all kinds of shortages.

As these new partnerships became more complicated, they also grew ever larger and heavier, no doubt finding it difficult to keep from sinking away from light or from moving too slowly to find food. Another cooperative solution—another transformation of imperialism into mutual aid—permitted them to move themselves farther and faster.

Remember that some breathers were shaped like twisting, lashing whips? Others had invented a proton motor. This was a spinning disk in a field of electrical potential, complete with microscopic ball bearings and an attached tail, or *flagellum*, so the bacterium could propel itself, literally motoring about! In the slime cities mentioned earlier, they can be seen along the canals among skyscrapers.

What might happen if these mobile bacteria stuck themselves onto larger cells in order to suck food from them? The big cells would find themselves moved about by lashing tails and propellers.

This is just what seems to have happened. The evolving cooperatives must have found the pushy bacteria worth feeding in return
for being driven to where they could find food or light. Eventually,
the propellers and flagellae evolved into shorter, stiffer *cilia*
arranged in rows so their rotating movements could be timed like
the oars of our ancient ships. Cilia were so successful that almost
every plant and animal cell living today has some part or parts that
evolved from these ancient rowing hairs.

But before these happy cooperatives had actually evolved, there
must have been a phase in which it was not clear whether cooperation would win out over competition, when the evolving protists
were like a factory of workers without management, or a world of
separate nations trying to take advantage of one another and all trying to give orders at once. After all, each kind of moneron had its
own DNA plans, its own welfare requirements.

What was called for was serious organization. And monera, as
we have seen, had already evolved a very effective communications
system that could make such organization possible—their own
WorldWideWeb. They knew how to trade DNA to revise their
own plans; they knew how to work from a common gene pool.
And apparently they drew on this experience to set up a new kind
of local gene pool, or information center.

Instead of collecting one another's genes inside themselves, they
streamlined themselves even further, giving up some of their DNA to
a common gene pool of general cell plans, which became the cell
nucleus. As time went on, the nucleus became a virtual library of
information for producing proteins and, in ways we still understand
poorly, probably took on the direction of the whole cell's affairs. The
individual monera of the new cell became less independent but
more secure, more inseparable parts of the new wholes.

This pattern of unity—>individuation—>conflict—>resolution—> cooperation—> unity-at-a-new-level-of-organization the ancient bacteria went through was mentioned and illustrated in Chapter 2, and is a typically repeating cycle in evolution. We humans are engaged in today, as we learn to cooperate at a global level, thereby achieving a new planet-wide unity.

Nuclear DNA, originally donated by many generations of participants in each of the evolving cooperative cells, continues even in modern cells to contain a tremendous amount of duplication. Many theories have been proposed to explain the repetitive nuclear genes, but perhaps there is no more reason for them than there is for all the duplicated documents in our own government offices. In any case, the huge quantity of nuclear DNA could hardly be kept in bacterial style as loose loops floating about in the cell. It would have gotten tangled and broken, messing up plans. Besides, the DNA might well have been destroyed by all the oxygen-making going on in these cells. And so it seems that the DNA was collected from each part of the cell, wrapped tightly in complex close loops around proteins, and stored within a protective nuclear membrane. This nucleus became a kind of central cell office for keeping DNA plans in order, but accessible for use, thus making possible better organization of all the cell's work.

This nuclear central office and library evolved into a staggeringly complex yet elegant organization we are still working to understand. The DNA in each of our cells, if stretched out, would be about six inches (or 15 centimeters) long, though it is packed into the invisibly small nucleus together with proteins and water. A jet plane, as Jeremy Narby has pointed out, flying one thousand kilometers per hour would have to fly over two full centuries to reach the end of all the DNA contained in a single human body's several trillion cells strung end to end!

What is even more remarkable is that a single handful of rich, natural soil, a large part of which is masses of bacteria, contains even more DNA than that because of the closer packing of DNA throughout their tiny bodies. That means that our entire planet is coated in DNA—just about the oldest surviving invention of all evolution—the language of life itself. Only now do we recognize DNA as intelligent in its own right, when we see it rearranging itself appropriately within organisms under stress. How foolish, then, are we humans, to kill and sterilize our soils of all life with chemicals, thinking we know better than nature how to engineer and grow our food.

When it was time for the first nucleated cells to divide, each of several very long twisted DNA molecules unfurled itself, then split and replicated itself as it had in bacteria. Theses pairs stayed buttoned together at one point along their length, by a *centromere*, meaning central place (also called a *kinetochore*, or moving spot), while each member coiled neatly around itself, that coil coiling into a shorter one, and so on until it formed a compact *chromosome* thirty thousand times shorter than the DNA molecule was when it started out!

Later we will describe the next steps in cell division. Here we just want to indicate the remarkable feats performed within this *nucleus*—this DNA-protein information center that evolved as the most important new feature of the giant cell cooperatives. Every living creature of Earth not a bacterium itself evolved from these nucleated cells, meaning that every living being of every kingdom of life beyond monera is made of the same basic kind of nucleated cell. Biologists call this superkingdom of cells, which includes the protist, fungus, plant, and animal kingdoms, *eukaryotes* (pro-nounced you-carry-oats, from the Greek meaning 'with a *karyon*, or kernel—the nucleus). To keep things straight, we call bacterial monera *prokaryotes* (pro-carry-oats, meaning 'before a nucleus').

In evolving these eukaryotic protist cells as bacterial coopera-tives, Gaia's creatures rediscovered some of the independence they had had before they became specialists depending on one another. The giant eukaryotes could now evolve new parts and ways of using them—all sorts of special membrane walls and internal skele-tons, gas-bubble vacuoles to control floating and sinking, other structures and chemical systems that helped do new jobs. A means was even found to circulate the jellylike cytoplasm in which all these structures are embedded, providing a transportation system for supplies and wastes.

Eukaryote cells, as we said, are on the average a thousand times bigger than prokaryotes, with a thousand times more DNA. They are in many ways as complex as human cities, or the bacterial colony cities described earlier. Until recently, scientists saw the nucleus as a computer behaving like an authoritarian dictatorship, containing all the information necessary to run the call and sending out 'top-down' command and control orders for what is to be built, produced, carried about, or otherwise done. Now it seems that the governing of cells is more decentralized—that the whole cell governs itself, using the nucleus as an information resource center.

Biology has made more progress in understanding the detailed composition of living things than in understanding their organi-zation as a whole, but this trend is now shifting. In our analogy with cities, some cell parts are structures like roads and buildings; others are chemical messengers, carrying instructions to and from the nucleus; some are production centers like factories; others perform services, taking in and delivering food, collecting waste, making repairs.

With ever more powerful microscopes, moving picture microscopy and animation techniques, we begin to understand how busy and lively, how complicated and amazing, life is inside such cells. But what we see leaves much to be learned about how it is all organized to function so smoothly. Even structurally we are still discovering things of major importance in cells. For example, until very recently we thought cells were bags of jelly-like stuff with the nucleus and organelles suspended within the jelly. Then, to their surprise, microbiologist Don Ingber discovered that cells have an internal architecture not unlike our bone and muscle systems— a *tensegrity* structure, with tension and compression components giving the cell integrity and the ability to move itself. Until his discovery, scientists had been dissolving this structure with the chemicals they used to prepare cells for study!

A huge new development is the understanding that nuclear DNA is reorganized in response to changes within and beyond the cell, that the entire cell, including its membrane or wall, is a creative autopoietic system.

All the cells of our own bodies (not counting the myriad bacteria living on and within us) are eukaryote cells, but we are just beginning to understand their evolution from ancient bacterial cooperatives. The story of this evolution has been simplified here. In actual fact it is one of the most fascinating and difficult puzzles ever to challenge biologists.

1993 Nobel Prize winning biologists Philip Sharp and Richard Roberts discovered that genes are broken into modules that can be reshuffled by *spliceosomes* referred to as a cell's 'editors' because they snip out inappropriate DNA sequences occurring between meaningful 'words.' Some of these modules are apparently shared by different genes, enabling evolution to proceed much faster than it could have if the old models held true. Especially interesting in light of this book's designation of a nucleus as a central library or information center is the new vocabulary of 'editors,' 'words,' etc. The picture emerging is consistent with the description given here of evolution as an intelligent process, rather than an accidental mechanical process. Sharp, in fact, speaks of a spliceosome as *knowing* where to cut and where to splice.

O   O   O

The DNA plans and composition of our own cells are, of course, unique to humans and differ from those of frog and fern cells, bacterial or Bactrian camel cells. When it became possible to analyze DNA in detail, it also became possible to identify a species by its own particular DNA pattern. Many species look very much alike, yet can be distinguished by differences in their DNA patterns.

Among the parts of our cells outside their nuclei are large numbers of tiny things that produce ATP energy currency. These cell parts have long been understood as little mechanisms for burning food molecules with oxygen to produce the cell's energy. But with rising interest in and understanding of DNA, biologists made a very strange discovery: that these little machines have their own DNA, the coded plans of which are quite different from those of the nuclear DNA.

How could cell parts be of a different species than the creature made of those cells?

Clues soon turned up. However different this DNA is from the nuclear DNA, it was found to be rather like some other DNA that biologists knew about—the DNA of bacteria quite like the breathers that evolved billions of years ago! At the time this was discovered, the story of bacterial cooperation in the evolution of eukaryotes was still unknown. Now that we know it, scientists are finding as many as a thousand different kinds of DNA outside the nucleus of a single cell.

The idea of *cell symbiosis*—the origin of eukaryotes as prokaryotes living together in cooperatives—had been proposed independently by a German, an American, and a Russian biologist around the turn of the century. All had noticed that the photosynthesizing *chloroplasts*—meaning 'green producers'—in the cells of plants resembled bluegreen bacteria. The Russian, K. S. Mereschovsky, suggested that other ancient bacteria had evolved into other cell parts. But biologists, who were trained to see living things as put together from mechanical parts, could not see cell parts as creatures in themselves.

Thus the symbiosis theory was ignored until Lynn Margulis an American microbiologist who became James Lovelock's partner in developing the Gaia hypothesis, revived it and produced a great deal of evidence to support it.

After much work, Margulis and others have shown that these energy-producing cell parts really are descendants of the ancient breather bacteria that came to live inside larger prokaryote cells, cooperating in building the first eukaryote cells. Luckily, teams of biologists working to unravel the ancient mysteries of cell symbiosis have found many clues in the behavior of today's bacteria. Rather vicious breathers can still be found drilling their way into other bacteria to reproduce there and eat the host bacteria from the inside. In the Tennessee laboratory of Kwang Jeon, protist hosts so invaded learned to tolerate and then to cooperate with their invaders in a mutually dependent relationship that brought about a new kind of creature. Surprisingly, this replay of the ancient evolutionary shift from outright aggression to full cooperation happened in only a few years' time.

Today, we find the descendants of the ancient breathers living and multiplying in the cells of every kind of protist, fungus, plant, and animal. It's high time we knew them by name. They are *mitochondria*—pronounced, mite-o-KON-dree-a—a word that comes from the Greek meaning 'thread grains,' because under a microscope they look like tiny grain hulls packed full of thread.

Using the oxygen we breathe, mitochondria make all the energy our bodies need to keep going and to repair themselves. Without our mitochondria we could not lift a finger. In fact, it is these swarms of ancestral bacteria, working night and day in all our cells, that keep us alive.

Or are *we* working for *them*? Lewis Thomas, in another of his perceptive and poetic insights, suggested that if anything in nature is a machine, perhaps it is us—maybe we are giant taxis which mitochondria built to travel around in safely and comfortably. Certainly mitochondria have done very well spreading themselves all over this planet, inside every other living thing, almost since the Earth came alive. There are so many of them swarming in our own cells that it's hard to guess at their actual numbers, but all together it is estimated they would weigh almost as much as the bodies they live in—that is, mitochondria make up much of our weight, and the weight of elephants and insects, clams and monkeys, toadstools and lizards, fish and worms.

In plants, from seaweed to sunflowers, potatoes, and palm trees, mitochondria live together with their relatives, the chloroplasts, which give plants their green color. You will easily recognize them as descendants of the ancient bluegreens and know that as they make energy from sunlight, water, and carbon dioxide, they also make the oxygen their mitochondrian cousins need for making *their* energy.

Mitochondria and chloroplasts, together with their still free-living monera cousins, the bacteria, are by far Gaia's most numerous and important creatures, though we are very late in recognizing our complete dependence on them. Quite to the contrary, we discovered bacteria in the context of medicine and treated them only as enemies. They are so hidden and tiny that, for years, we paid no attention to their good works, but we now know that their cooperation is the essence of the entire Gaian life system.

Besides making the vital gases oxygen and carbon dioxide for each other, the chloroplasts and mitochondria of eukaryotes, together with their prokaryote cousins, form other cooperative cycles, for example, the food chains mentioned earlier. While plants make their organic bodies from simple minerals, water, and carbon dioxide, animals can only make their bodies from the ready-made organic molecules of plants and other creatures. It is primarily bacteria that cause the decay of dead plants and animals, reducing them to the simple substances on which new plants can live.

Margulis' discovery, that eukaryote protists evolved cooperative internal schemes to overcome the problems caused by competition among prokaryote bacteria, was almost as much a shock to the world of science as was the Gaia hypothesis itself. Besides showing that cell 'mechanisms' such as mitochondria are creatures in their own right, she was suggesting that harmonious cooperation played a big role in evolution. This ran counter to the beliefs stemming from Darwin's work, adopted by scientists in western countries, that evolution was just a survival race driven by competition.

The theory of evolution through competition has played a big role in our world, not only in science but in shaping our whole human outlook and way of life. Only by understanding its origin and its widespread effect on our lives can we understand how to change our view of ourselves and our role within our larger Gaian planet. Let's look, then, into the way our Darwinian view of evolution came about.

7

Evidence of Evolution

Charles Darwin was an English gentleman scientist who lived during most of the nineteenth century and who traveled far and wide to study nature. Seeing the great variety of plants and animals around the world, he was struck by the way each kind seemed uniquely suited to the place in which it lived. He believed, as had a few other scientists, that all these living things must have changed over time to suit their particular environments so well. This was a novel idea at a time when almost everyone believed that the world itself was unchanging and that God had created all species at the same time, just as they are. How could they have changed? Despite Darwin's careful observations and ideas of likely changes, he had no theory, no way to explain how they could occur.

Nature, he noticed, produced great numbers of seeds and eggs for all kinds of plants and animals, though only few of each kind grew up. Somehow it seemed that only those best suited for survival in their environments—the fittest of each generation—grew up to reproduce offspring like themselves. But how could nature recognize and choose them?

Darwin had seen plant and animal breeders choose the fattest grains or the fastest horses of each generation to be the parents of the next generation. This was possible because, in each species, the young were not all exactly alike, but were as varied as human brothers and sisters. By such selection, generation after generation, the breeders changed the species, producing ever fatter grains or faster horses. This was exactly the kind of thing nature seemed to be doing, but it still puzzled him how nature selected the fittest creatures of each generation.

The theory of evolution finally came to him when he read an article about food shortages and starvation. Could it be that all nature's young had to compete for food when there wasn't enough for all? Was this how nature put creatures to the test? If so, then surely the fittest in this competition would pass the test and survive to grow up.

The fittest bears, for instance, would be the ones with heavy coats to keep them warm as they hunted for food in cold places; in warm places bears with lighter coats would more likely survive. Birds with long beaks were the fittest where worms had to be pulled out of holes; birds with short, strong beaks would survive where seeds had to be cracked. The bears or birds born with the fittest coats or beaks would win the struggle for food and grow up to produce babies while others starved or froze or were eaten. Their babies would inherit varying degrees of 'fit' coats or beaks, and again the fittest of all would be selected in competition.

Everything seemed clear now. Large numbers in the face of too little food produced competition, and competition led to *natural selection.* The selection itself was based on the natural variety of creatures, some of which were fitter than others.

In Darwin's time, of course, environments—such as deserts or mountaintops or sea bottoms—were seen as *places* in which living things made their living, not as live parts of a great living planet. Each environment would select for different kinds of fitness in the competition for food, for mates, for the best hiding or nesting places, and so on.

If the wind scattered the seeds of the same plant into different places, a desert would select for young plants that could live on the least water, while a windy mountaintop might select for those with the strongest grip on rocks. Just so, some thick-coated wolves that could do well in cold environments might wander to warmer places and die out while wolves with thinner coats survived.

Offspring of the same original parents might therefore become quite different over many generations as a result of settling in different environments that selected for different body patterns or features. When they became so different that they could no longer mate with each other, they had become different species. Thus Darwin explained the evolution of new species.

In Darwin's theory, then, unexplained accidents of birth that made creatures fit better into their environment were selected for survival and passed on to future generations. Unlucky accidents that made creatures less fit were rejected by natural selection and died out.

Scientists quickly made Darwinian evolution fit the idea of nature-as-mechanism by regarding creatures more or less as wheels fitting the cogs of other wheels in the great clockwork of nature. Some wheels just happened to be made better than others by lucky mechanical accidents during their replacement, or reproduction.

The idea of natural competition leading to the survival of the fittest appealed to men who were obsessed with the new social structure of industrial capitalism.

With the advent of genetics, accidents of birth were discovered to reflect changes in genes. When the structure of DNA and its copy process were understood, these accidents were believed to occur on a random basis—meaning without any pattern—as DNA copied itself or was damaged. Most biologists today still see accidents, now known to occur in DNA, as the only source of natural variation, despite growing evidence that such accidents are detected and repaired very quickly.

Ever since Darwin, our general view of evolution has been of a battle among individual creatures pitted against one another in competition for inadequate food supplies. Only now are we in a position to understand the Earth as a whole—a single geobiological dance woven of many changing dancers in complex patterns of interaction and mutual transformation.

Competition and cooperation can *both* be seen within and among species as they improvise and evolve, unbalance and rebalance the dance. Consider again the spiraling pattern described as unity—>individuation—>competition—>conflict—>negotiation—> resolution—>cooperation—>new levels of unity, and so on. Note that competition and cooperation are different phases of the cycle. Young species tend to grab territory and resources, maximizing the numbers of their offspring to spread themselves where they can. As species encounter each other, conflict develops in the competition for space and resources. Eventually negotiations leading to cooperation prove useful to the competing species and they reach the higher level of unity, as we saw happening in the transformation of monera into protists.

Evolution is this improvised dance of transformation in which ecological balance is worked out again and again. Remember that living things *have* to change, even to stay the same. They have to renew themselves *and* adjust to the changes around them. Rabbits evolve together with their habitats, and we might call that the dance of rhabitats! All creatures evolve in concert or connection with all else evolving around them. New levels reached in the unity spiral through phases of competition and cooperation are examples of what we described as mutual consistency. The internal harmony of our evolved multicelled bodies is a good example, but our global society is not, as it is still struggling to get beyond its competitive phase.

It took a century and more after Darwin's theory was published for us to understand that environments are not ready-made places that force their inhabitants to adapt to them, but ecosystems created *by* living things *for* living things. All the living things belonging to an ecosystem, from tiny bacteria to the largest plants and animals, are constantly at work balancing their lives with one another as they transform and recycle the materials of the Earth's crust.

Darwin, along with Lamarck, and Wallace, were modern pioneers in showing us that species evolve and attempting explanations of how this could happen. Their theories were a great step forward for science, since religion had put an end to all theorizing about evolution since a few ancient Greek philosophers, such as Anaximander, had thought about it. Anaximander had said that everything forming in nature incurs a debt, which it must repay by dissolving so other things could form—a marvelous description of evolution through recycling in a single sentence!

Now we can see that Darwinism—and its updated version, neo-Darwinism—is a misleading way of seeing nature. The notion of the separateness of each creature, competing with others in its struggle to survive, had well described, and justified, English and American societies' new forms of competitive and exploitative industrial production in a world of scarcity. But we are now beginning to understand that humans must learn to harmonize our ways with those of the rest of nature instead of exploiting it and one another ruthlessly. The social view of individual people pitted against one another in such struggle makes little more sense as an ideal than the notion that our bodies' cells are competing with one another to survive in hostile bodies. It is simply no longer useful or productive to see ourselves as forced to compete with one another to survive in a hostile society, surrounded by hostile nature.

The point here is that we *do* see ourselves in such competition, not because this is inevitable, but because Western science developed in close harmony with social and political traditions that welcomed these ideas. The Darwinian theory of evolution was applied to forming a society, a social system, designed in accord with, and justified by, the Darwinian concept of nature. If we learn to see evolution as a single holarchy of holons working out the mutual consistency of cooperative health and opportunity, we can set up a social system to match that view.

History may someday record the greatest discovery of twentieth-century science not as nuclear power or electronics, but as the recognition that there is no absolute truth to be discovered about the world—that scientific theories can be judged only by their usefulness to science and ultimately to all society. Definitions of usefulness often change over time, and thus scientific 'truths' must necessarily evolve along with human society.

Neo-Darwinism insists that random accident and natural selection are the sole 'mechanisms' of evolution. Yet the self-organized creatures and ecosystems—rhabitats—such as that which we saw evolving through the genetic information exchange web of bacteria, including their negotiated organization of nucleated cells are not readily explained as simple accumulations of lucky accidents. Nor does natural selection amount to a real theory, since it tells us little more than that some creatures die before they reach the age of reproduction. A modern theory of evolution must concern itself with the way in which natural holons are organized and maintained in holarchies, with descriptions of continual interactions among the levels of DNA, organisms and whole ecosystems. It must also deal with the aesthetics of orchids and butterfly wings and dolphins creating bubble rings and other games for the pure joy of it.

O   O   O

As pointed out earlier, no place on Earth today, not even the barest-looking mountaintop or the deepest part of the sea, has fewer than a thousand different species from various kingdoms— monera, protists, fungi, plants, and animals. Yet what we humans see as living things are only the largest of the plants and animals— beings the size of bugs and bushes and beluga whales, creatures on our own size scale. The vast majority of Earth creatures, however, continue to be microscopic monera and protists. Think once again of our rocky planet rearranging itself through chemical activity into a rich network of bacteria and environments that are good homes for bacteria. This is what most of the activity of our living planet is still all about.

The age of protists was a long age of creating amazing diversity in the larger single-celled body forms and lifestyles before multi-celled creatures arose. As Gaian evolution continued, ecosystem holarchies went on transforming themselves, as we will see, into ever larger living bodies grown from single cells: worms and insects, fishes and amphibians, funguses, flowers and trees, reptiles, birds, and mammals. Still, the smallest living creatures are even now those that work hardest to create the environments needed to sustain the larger plants and animals. For that matter, larger creatures are really extensions of the microcosm. As eukaryote cells are evolved from prokaryote cooperatives, multi-celled creatures are later cloned upon them, as we will see.

If we think of every ecosystem holon as a kind of body, we can regard each creature as a cell, each species as an organ with a unique function. Their holon evolves its ecological balance as a whole. We really should talk about co-evolution, rather than evolution, to remind ourselves that no species can or does evolve by itself, but that all must cooperate by adapting to, or negotiating with, the others' steps in the dance of life. Thus they reach mutual consistency with one another and with the rest of their surround. The story of co-evolution is still being put together as we try to understand that creatures are not just passive mechanical cogs in a wheel but active agents in their own evolution, or co-evolutionary process.

O   O   O

To some extent we can study the co-evolution of species within their present ecosystems and find clues to how they must have co-evolved in the dim past, but we also have many other clues to life's long-ago patterns. Fossils—the imprints and remains of creatures turned to rock—are very important clues to what early creatures looked like, especially now that we know how to tell their age and have found fossils of even the most ancient bacteria in rocks, such as the stromatolites they built billions of years ago. Fossils alone cannot prove that one species changed into another, but the fossil record clearly shows that larger, more complicated creatures existed only after smaller, simpler ones had paved the way for their existence.

In Darwin's theory, species change very slowly but steadily as environments select for the tiny accidental changes that make some individuals of each generation slightly fitter than others. Since his time, however, many more fossils have been collected, and we see that most species must have changed in spurts from time to time—far more quickly than such accidents can explain—while others have scarcely changed at all, despite the slow, steady stream of accidents they must have endured.

The known rate of DNA accidents seems to have no relationship to the rate of change in creature patterns. Some bacteria living now are like those that lived billions of years ago. Squid and sharks and ants have stayed much as they were hundreds of millions of years ago, as we see clearly in the fossil record. Yet many of these slow-to-change creatures' fellow species in co-evolution, such as ourselves, have become very different modern species.

Species whose genetic plans have hardly changed are like bicycles in a world of jet planes—they still work very well as they are, yet they have been steps along the way to bigger, more complicated inventions. They are, in a way, living fossils to compare with creatures that apparently continued to change or evolve. It seems that co-evolution has rhythms like any other dance—some slow, some so fast in comparison that they seem almost to leap from being one kind of creature to being another. We ourselves are a good example of a species that changed very rapidly.

If all living beings were created at once—as creationists still believe—then all modern species should have fossil ancestors quite like themselves. But this turns out to be true of only very few creatures larger than monera. There are estimated to be somewhere between three million and ten million species alive today, yet well over 99 percent of all the species that ever lived are now extinct. It is worth noting here that in the times of most rapid extinction it is estimated that the rate was about one species lost every thousand years, while we humans have probably caused the extinction of a million species in the last quarter of this century alone! Many biologists acknowledge the present as the sixth great extinction, as we pointed out earlier.

Now and then fossils give us wonderful clues to change, such as those showing reptiles evolving into birds. The *archaeopteryx*—Greek for 'ancient-wings'—was a kind of flying reptile with a horny beak and wings strong enough to propel its big body through the air. Its skeleton looks very much like those of modern birds, and the fossil imprints are so clear we can almost see reptile scales evolving into feathers, front legs into wings, and long snouts into hard beaks. The babies of birds, even today, still hatch from eggs just as their reptile ancestors did.

The fossil and chemical geological records show that dinosaurs and other large creatures of their time died out around sixty million years ago during a period when there were major changes in their environment. Apparently these large creatures could not adjust to big changes in climate caused, it now seems, by the impact of a giant meteor or planetoid. Smaller animals did survive the crisis, but they must have evolved so quickly that none of them left clear fossil records of the gradual changes inferred by Darwinian evolution.

The fossil record alone is simply not adequate to prove modern creatures' lines of descent. In fact, scientists have not found a single clear and complete fossil line of descent for any modern creature, and this has become a major argument used by the creationists, who don't believe in evolution. How can we be so sure that modern creatures actually descended from earlier ones that were quite unlike them?

We have, fortunately, other clues to evolution. For one thing, we can actually *see* evolutionary changes taking place in some living species. In bacteria, protists, fungi, plants, and animals that reproduce quickly and often, we can follow the changes over many generations. With electron microscopes and other modern instruments we can actually track their changing patterns of DNA as well as the more obvious changes in cell or body structure that follow from the microscopic changes.

It was of course a big surprise for biologists to find that creatures can rearrange their own DNA in ways that can hardly be called accidental. In using antibiotic drugs to kill particular kinds of bacteria, as mentioned earlier, we often find that a species we attacked successfully has suddenly changed into a species that cannot be harmed by the drug. We say it has become resistant. But such new resistance implies genetic change—the kind of genetic change we now know all bacteria to be capable of through their WorldWideWeb of DNA recombination—the system that helped them cope billions of years ago with such life-threatening matters as ultraviolet radiation and poisonous oxygen in their environment.

Individuals in bacterial colonies change their DNA rapidly and effectively when the colonies are deprived of their usual foods and forced to live on what they were previously unable to digest, as Ben Jacob has shown. Even more impressive is the fact that larger, more complex creatures also can change their DNA in emergencies. Many insects and plants that we humans consider pests have made themselves resistant to our chemical poisons in just a few generations. Recent work in biology, such as that of Mae-Wan Ho, shows that instructions do not flow only one way from nuclear DNA to bodies, but that bodies can register changes in the environment and in themselves, communicating these changes to their DNA in ways that offspring may inherit.

Many other examples of intelligent and specific genetic change in response to circumstances outside organisms have been shown over the last half century. Just as we are finding that organisms are not separate from their environment, we are also finding that DNA is not separate from *its* environment within and without its organism. We know now that the DNA recombination capabilities of cells include the ability to reproduce DNA between cell divisions—that genes can be copied and relocated when necessary. And this is only the beginning of our understanding of how organisms change their plans in order to survive in health.

The idea that the environment can trigger inheritable changes in creatures had actually been proposed before Darwin's opposing idea that environments can only select among changes produced independently of environments. Jean Lamarck, who named the study of living things *biology*, proposed it as the first modern theory of evolution. Lamarck's explanation of how living things changed themselves—the classic example being giraffes stretching their necks to reach higher leaves and then passing on long necks to their offspring—was not as convincing to Western scientists as Darwin's theory of accidental variety and natural selection through competition. Thus Lamarck's theory was ridiculed while Darwin's was adopted in the West, though Lamarck continued to be respected in Russia.

Neither Lamarck nor Darwin, of course, knew about DNA or even had a theory of genes, so they could not even guess that bacteria can trade around their genetic material or that larger multicelled

creatures can rearrange their genetic material on the basis of experience in their environment. For that matter, many biologists still
resist accepting the new evidence consistent with Lamarck's theory.

There is still a great deal to learn about biological information
systems. Earlier we mentioned that the role of most nuclear DNA
is still unknown and that it contains many extra copies of some
genes. We guessed that the excess may be a hangover from the time
when many bacteria contributed to the nuclear DNA. But it may
also be, as some biologists believe, that all through evolution
species collect and pass on reserve genes, which may someday be
useful in an emergency. After all, we keep libraries filled with
books, the vast majority of which are unused at any given time.

It is certainly obvious by now that DNA can reorganize itself
and repair the kind of accidental change that was thought to be the
only way to evolution. It is a relief to know we are not just a lot of
piled-up accidents and copying mistakes, but beings who have
organized and evolved ourselves in harmony with the other living
beings that form our environment. It is good to know that life is
too intelligent to proceed by accident!

Environments simply are not fixed places that living things must
fit into, adapt to, as we had thought, but the busy activity of living
things themselves, working out their ways of life together as parts
of the live Earth. This co-evolution in ecosystems now seems a
matter of creatures changing themselves from the inside, in
response to their environment, and *also* prodding changes in other
creatures that are part of their environmental holon. A change in
one species will thus be reflected by changes in some others. The
Unity cycle reflects these 'negotiations' in which living things
evolve themselves and are evolved by one another, together evolving mature balanced ecosystems—such as rainforests—from

immature ones with relatively few initially-competitive species. Note that each species is continually incorporating raw materials into its bodies, and being transformed in turn into raw materials for others, as Anaximander observed.

We only began studying *ecology* a few decades ago, when we recognized things going wrong in our environment because of changes we were making, especially after Rachel Carson called our attention to this enormous problem. We had been creating erosion and deserts by cutting down forests, poisoning land and sea creatures all over the planet with our pesticides and herbicides, polluting and destroying our air, water and soil with various chemicals, creating monocultures and warming the climate unnaturally.

Suddenly we began to study our planet's ecologically balanced body with attention to ecological 'illnesses' and their possible cures, just as we had begun earlier to study our own bodies with attention to what went wrong with them—to *their* illnesses. As Lovelock has pointed out, medicine began to make real progress only when we began studying the physiology of our bodies—the way their interwoven systems work under normal, healthful conditions—and so we will make more progress in understanding ecology and evolution as we learn how our planet's normal physiology works. How are its many kinds of supplies recycled? How do its information systems work to adjust imbalances? How do its countless and varied creatures contribute to its overall health?

O   O   O

Much groundwork for planetary physiology can be found in the work of Vernadsky, the Russian biologist who described life as a disperse of rock, or rock rearranging itself, as we have been calling the same concept. Vernadsky pointed out that living organisms were originally built of the inorganic minerals of the Earth's crust and still contain such inorganic minerals, that they transform such inorganic minerals into living matter and living matter back into inorganic minerals. For this reason he saw no separation between biology and geology, but became interested in the constant transformation going on from the one domain to the other.

His concept of all living creatures together as living matter was used to propose that *their* part of the Earth's crust is energetic enough to actively transform the more passive parts—what we have called the geological parts—into itself and its products, literally *feeding* on it. On the surface, this concept of living matter is the same as Lovelock's concept of biota, as the sum total of living creatures, contrasted with their abiotic, or nonliving environment. But in Vernadsky's conception the emphasis is on geological continuity—on each part as a transformation of the other—whereas in Lovelock's conception the emphasis is on their interaction as separate parts of a working system. Oddly, Vernadsky, who apparently did not see the planet alive as a whole, perceived its integrity more fundamentally than Lovelock, who does see it as alive. Vernadsky was interested in the fact that the *same* atoms over time would alternate as part of animate and inanimate matter.

The processes by which organisms build up and destroy their own bodies—their particular structures built of proteins, water, carbon compounds, and minerals—is called *metabolism*, from the Greek word for change. Metabolism, which we touched on earlier in discussing the law of entropy, is a process of chemical changes in living matter by which energy is provided for taking in new matter, building and repairing cells, collecting and excreting wastes. Metabolism is divided into two parts: *anabolism* and *catabolism*, the buildup and breakdown of body substance, or *protoplasm*.

Metabolism, then, is the most basic autopoietic activity of all life. It recycles the materials of the Earth's crust into animate matter and then back into inanimate matter that can be used again to create more living matter. Vernadsky understood metabolism as the activity of all Earth's living matter taken together, as well as that of any particular organism, since he saw all living matter as a constantly shifting high-energy portion of the Earth's crust. We earlier observed that virtually all of the Earth's atmosphere, seas, soil, and rock are made from the products and dead bodies of organisms.

Even the hardest pure-carbon diamonds are pressed coal, which earlier was pressed animal and plant bodies. The sedimentary rock formed by pressure on the ocean floor, as another example, begins as sediment, including vast quantities of algae and animal shells, all passed through the guts of sand and mud-eating worms to further transform them, just as soil is transformed by the related Earth-eating Earthworms of dry land. Life, then, is the most powerful of geological transformers. That the record of Earthlife's evolution lies everywhere in geology, not only in recognized fossils, is referred to in the title of a book on Vernadsky's work, called *Traces of Bygone Biospheres*.

The geological activity of creatures also includes their production of atmospheric gases and their transfer of groundwater back into the atmosphere, a process that is clearly visible in the pumping action of rain forests, the rain then falling to dissolve more earth and rock. On the whole, however, the geological activity of creatures is less the larger they are, most of this work being done by microbes and rock- or mud- or earth-eating worms.

Some microorganisms contain half a million to a million times as much of some mineral, such as iron, manganese, or silver, as their environment does. The concentration of elements is another way in which life alters Earth's crust. Microorganisms are responsible for concentrating the radioactive materials, such as/ uranium, that we mine to produce atomic energy—perhaps concentrating it in their habitats to keep themselves warm! Copper and other metals we mine have similar origins. From metal veins to continental shelves and the entire atmosphere, not to mention the composition of seas and soils, we see the staggering work of Earth's microbes.

It was so clear to Vernadsky that the activity of living matter was metabolic that he proposed we reclassify living organisms on the basis of their metabolism. He argued that our present classification from kingdom to species by way of phylum, class, order, family, and genus had led us to classify as related organisms many that really are not related under natural conditions. A better scheme, he felt, would be to divide kingdoms according to the way in which each of their species metabolizes supplies from its environment. The different ways in which organisms feed themselves had already been named by the German biologist Wilhelm Pfeffer. Vernadsky proposed them as a biological classification scheme.

In this scheme the metabolic process of organisms begins with the category called *autotrophs*—self-feeding organisms that can build their own giant molecules, such as protein and nucleic acids, from simple molecules and elements such as minerals, water, and carbon dioxide. This category includes the *photoautotrophs*—self-feeders that use sunlight in metabolizing basic molecules. A second major category of organisms is called *heterotrophs*—meaning feeding off others, because its members cannot make large molecules from basic ones but must eat other organisms for their ready-made large molecules. A third category is called *saprotrophs*—meaning to feed on the dead, because its members eat dead bodies and reduce their large molecules back to the basic ones the autotrophs can use.

The fourth category is *mixotrophs,* which can metabolize in more than one way.

Finer distinctions within these categories are made as heterotrophs feed off other heterotrophs, and so on. What is important about this scheme is that organisms are classified not by their structures but by their functions within the whole geobiological life process. It recognizes organisms as self-organizing packets of the Earth's crust with enough energy to move about the more sluggish matter around them. Vernadsky even suggested that evolution might proceed by the natural selection of organisms which most increase *biogenic,* life-originated, energy—the energy to move around the atoms and molecules of the Earth's crust at the highest speed.

The energy of living matter sometimes explodes almost beyond belief. A locust plague of a single day has been estimated to fill six thousand cubic kilometers of space and weigh forty-five million tons! It is the locusts' heterotrophic metabolism, of course, that makes them a plague as they suddenly convert vast quantities of the autotrophic crops planted by humans into their bodies. Most bio-geologic activity goes on less dramatically, though it is impressive enough to consider that a single caterpillar may eat two hundred times its weight each day. All ecological areas have more autotrophs than heterotrophs—it takes more of the former to sustain the latter. Thus, a forest may have 2,000 to 5,000 times as much autotrophic as heterotrophic living matter.

Vernadsky did not consider this classification scheme the only one possible; he recognized that one could learn much by trying other methods. One of his schemes was based on the type and amount of mineral content in organisms. Certainly he was one of the first modern scientists to see the Earth in a truly holistic way and to provide evidence of its evolution through the transformation of rock into living creatures and back into rock.

Many scientists have since built on his work or developed independent studies of the Earth from a holistic perspective, but Vernadsky's work has been given particular attention here because its fundamental conception of biogeochemical unity is so important and so little known in the West. Our best western scientific progress in understanding Gaian physiology has been through the work of Lovelock, Margulis and their co-workers.

8

From Protists to Polyps

If you look at a drop of pond water under a microscope, you will see a great variety of protists living together, creatures you would never have suspected were there, unless you've done this before. You may see, for instance, *paramecia*, tiny slipper-shaped algae rowed along by lineups of waving cilia oars. Then along will come a giant, blobby amoeba, changing shape before your eyes as it pushes out *pseudopods*, 'false feet,' in its search for food. Perhaps it will get stuck in a tangle of rod-like algae, strings of cells sitting quite still and making energy from light with their green chloroplasts. Other green algae, such as *euglena* with their long whip-like tails, may also flit by.

If you look instead at a drop of seawater, you may see whirling *dinoflagellates* that glow in the dark by making their own light or *diatoms* and *radiolaria*, looking like beautiful crowns or glass ornaments. There seems to be no end to the fantastic patterns of these tiny single-celled creatures, whether they live alone or stuck together in colonies like the ball-shaped *volvox*.

The world of single-celled creatures in a drop of water is probably much as it was a billion years ago when there were no larger creatures. Yet these same protists went on to build and evolve all the larger multicelled creatures of Earth

On today's oceans great blankets of plankton, mostly made of protists, float about ever renewing the atmosphere with nitrogen, oxygen, methane, and other gases they release. Sulfur dioxide produced by plankton actually seeds the water droplets forming clouds. This means that the cloud cover over much of the planet, and thus the planet's warming and cooling system, is regulated in large measure by these tiny creatures. They also provide food for many heterotrophic species, from the almost microscopic shrimp that swim among them to the largest of whales. The wastes and dead bodies of those that eat plankton sink to the bottom where myriad saprotrophs decay them back into molecules that may come to the surface to nourish new plankton.

Plankton not only serve as part of a food cycle but play most important roles in balancing the chemistry of the atmosphere and the seas, as well as in the geobiology we talked about in the last chapter—the transformation of the Earth's crust from rock to living matter and back to rock.

Among the creatures forming plankton are thousands of species of diatoms and radiolaria, those particularly beautiful protists whose shells look like fancy glass ornaments when seen through a microscope. In Gaia's dance there are more diatoms than any other kind of creature except bacteria and they are responsible for transforming a great deal of rock.

Land rock, we know, is dissolved by streams and rivers into salts and minerals, which they carry to the sea. Life in the sea needs these building materials, but like the gases of the air, they must be kept in balance. If too many of them pile up in the sea, living creatures will choke. One such mineral is silicon, in its silica form (silicon dioxide). A huge amount of silica is washed into the seas every year—hundreds of millions of tons of it. But huge numbers of diatoms wait in the sea for these silica supplies, for silica is just what they need to build their sparkling shells. When they die, the diatoms sink to the bottom, leaving their silica shells to settle into rock—three hundred million tons of silica rock every year!

The number of diatoms in the sea naturally adjusts itself to the supply of silica brought by the rivers. There is always just the right number of diatoms to use up the silica dissolved in the sea. This well-balanced system—water dissolving rock, diatoms and other protists making it into lovely silica shells, then being eaten by others or dying so their shells sink down to the sea bottom to become new rock—worked itself out in co-evolution as a kind of transport system within the great Gaian body. The shells of radiolaria are also part of this process. The famous white cliffs of Dover on the English coast are ancient deposits of the microscopic, chalky, snaillike shells of yet another marine protist—*foraminifera*—which were pushed out of the sea by the endless motions of the Earth's crust.

Many other geobiology cycles or systems are still to be discovered as we keep working on the puzzles of planet physiology. Unfortunately, our old view of nature made us see ourselves as just one of the many creatures competing for survival on a planet without enough for all to live. But now that we have microscopes, telescopes, spaceships, computers, and other instruments that show us so much more than we could see with our eyes alone, we are in a position to understand the pattern of life within life from the largest to the smallest holons.

We know the mitochondria cooperating in our cells long ago worked out a mutually consistent way of life with other cell parts. We know they and we have a mutually consistent arrangement as we provide their fuel and safety while they make our energy. We can see how all species, including our own, must work out their mutual consistency with one another as co-evolving parts of the great Gaian body.

O   O   O

Of all the cooperative steps in Gaia's dance, we saw that one of the most important was the invention of sex—the sharing of creature plans by uniting DNA from more than one source to create a new being. Because the exchange of genes among bacteria worldwide was and is so free and continual, biologists had to give up their attempts to classify species of bacteria—recall that species are identified by their DNA. We can only identify strains of them that keep recognizable forms despite the free trade. Remember that this kind of original sex had nothing to do with reproduction, as Margulis pointed out in tracing the origins of sex. It is because of this sexual freedom, this efficient communications system lost forever in later kingdoms of life, that bacteria could remain streamlined creatures with tremendous flexibility, able to trade information worldwide and thus solve almost any emergency situation.

In the kingdom of protists, sex took some strange new twists, very likely quite by accident. These twists eventually linked and limited their sex to reproduction and to two partners within the same species. Thus the boundaries of sexual reproduction became our way of defining species boundaries. The within-species sex of the protist kingdom was passed on to multicelled creatures though sometimes different species co-evolved to help each other in their reproduction, as in the case of flowering plants cross-pollinated by birds, bats, moths, bees, or other insects. But before we get to larger creatures, let's see just how the kind of sex we know—the production of offspring by the mating of two parents—came about among protists.

A prokaryote without a nucleus, remember, reproduces by splitting or budding after its loose DNA loop unzips and copies itself. Even though this process gets a bit more complicated with the great quantity of DNA packed into a eukaryote's nucleus, most eukaryotes, including our own body cells, also divide in this way. The process begins with the unzipping and copying of each section of DNA, their neat coiling and recoiling into chromosome pairs buttoned together by centromeres or kinetochores, as we saw in Chapter 6. The nuclear wall dissolves to release them, and the cell constructs a fabric of microtubules on which the pairs of doubled chromosomes line up. These then unbutton themselves by dividing their kinetochores, which then button themselves to the tubes such that the two members of each pair of doubled chromosomes ride smoothly in opposite directions, pulled by the kinetochores to opposite ends of the cell before it divides into two with a full set of chromosomes each.

This usual way for cells to divide is called *mitosis—mito*, as in mitochondria, meaning thread, such as humans invented for weaving. If you could watch cell mitosis through an electron microscope, the neat formation of microtubules called the mitotic spindle and the shuttling of chromosomes along its threads would indeed look like a weaving process.

Mitosis is a non-sexual, or asexual, way to reproduce. The offspring of mitosis are clones—offspring of a single parent cell usually thought of as being exact copies of that parent. Of course we know that our whole bodies are cloned from the single fertilized egg we began as, and our cells are very varied. The idea that clones are all alike is linked to the idea that sexual reproduction is what brought variety into evolution. But the clones of rare asexually reproducing animal species, such as certain lizards, are quite as varied as the off-spring of the more usual sexually reproducing species.

If sexual reproduction did not evolve by natural selection for the advantage of variety, as scientists thought for so long, then why *did* it evolve?

Again we owe the tracing of a story from the ancient microcosmos to Lynn Margulis and her co-workers, who combined clues from earlier biologists and from their own research. Margulis noted that sexual reproduction has three important aspects: the halving of chromosome numbers within each parent, the doubling of chromosome numbers by bringing two parent cells together, and the alternation between these two stages of halved and doubled numbers generation after generation.

How odd to halve chromosomes continually, only to double them again. Margulis investigated this mystery by looking for cases of halving and doubling chromosomes in the microcosm and for ways in which they might have become linked into a single reproduction system. The story that emerged is, like the evolution of eukaryote cells, one that begins with exploitation and ends in cooperative partnership, and once again starvation is the initial motive.

A desperately hungry protist, even today, may resort to cannibalism, and on occasion may fuse the swallowed victim's nucleus with its own. All nucleated cells will fuse with one another under the right conditions. Doubled chromosomes may also come about when a protist begins mitotic division and is then unable, for some reason, to finish the process, failing to divide after doubling the number of its chromosomes and fusing them back into a single nucleus. This, too, has been observed.

In either case, the extra chromosomes may work well in times of need but become unwanted extra baggage when things go well. So protists learned long ago—over a billion years ago, which was not so long after they had become protists—to reduce the number of chromosomes again when this was to their advantage. The process of halving a cell's chromosomes is called *meiosis*, which means 'lessening.' Some protists seem to have become experts at doubling and halving their chromosomes according to the demand of changing conditions from drought to plenty and back.

The *fusion* of two sets of chromosomes into a single nucleus—if they are from different protists, even if one protist has eaten the other—is a sexual union. The *halving* of chromosomes in meiosis, as we just saw, was a solution to the unnecessary and troublesome burden of doubled or tripled chromosomes. What we call the haploid, or 'half-set,' of chromosomes that seeds, pollen grains, eggs, and sperm contain, are half of our normally double, or diploid, number.

The chromosomes of all our body cells are paired, one of each pair from each of our parents. Far back in evolution this doubling must have occurred as described and stuck as the normal number. When sexual nuclear fusion became linked to the reproductive formation of new generations of individuals, the double number had to be halved before each sexual-fusion and reproduction event to avoid doubling the chromosomes mercilessly in each generation, which would have been literally a dead end. Sperm, pollen, and egg cells are all produced by this meiotic halving process in such a way that the fusion of egg and sperm or egg and pollen results in the normal diploid chromosome sets of animals and plants.

Cannibalism, fusing nuclei and then reducing chromosomes again, accidents of timing, and perhaps other events finally worked themselves out into a reliable system of sexual reproduction—not the most elegant system nature ever devised, but one that has obviously worked well enough in the world of creatures less elegant and less sexually free than bacteria.

To carry out the work of a developing embryo, the DNA of haploid chromosomes from two parents must match stretch for stretch, gene for gene. As new species branch away from each others DNA sequences, their offspring may be infertile, as in the case of mules born to horses and donkeys, or tiglons born to lions and tigers. With further separation, mating become unproductive and ceases altogether. Branching species usually branch by occupying different ecological areas and so do not normally find each other to attempt mating, but humans have shown the possibility, though their hybrid offspring are sterile.

The change to the new way of reproducing did not happen all at once. Even today, in fact, some protists, such as paramecia, still reproduce both in the old way of mitotic cloning and in the newer way of sexual reproduction. Paramecia are a good example of nature's experiments with sex and reproduction, having as many as eight different sexes rather than only two, if we so want to label mating or gender types.

Many protists can reproduce with or without sex, that is, sexually or asexually. Sometimes one way serves their needs better, sometimes the other. Diatoms, for example—those lovely tiny creatures with the fancy silica shells—tend to reproduce just by mitotic splitting. They manage this by making their shells in two pieces that can come apart, one piece slightly larger, fitting neatly over the edge of the other to close it, just like a round pillbox with a lid.

The trouble is that while all the offspring must complete the half-shell box they inherit, diatoms have learned only how to make the smaller bottoms of box shells when given half a shell to start with. That means the split-off diatom that gets the bottom of the box has to use it as a top, making a smaller bottom than its parent diatom had. Over generations, then, lines of offspring inheriting bottom shells get smaller and smaller— only those with direct inheritance of the original top shell maintaining the original size. This problem is eventually solved by sex!

Instead of continuing to split by mitosis, they resort to meiosis, producing little packages of single-set chromosomes called *gametes*, or sex cells, that have no shells at all and behave the way eggs and sperm do. When two of these gametes get together, it seems they have *all* the DNA plans for a new diatom, including plans for the top *and* bottom parts of a normal-size diatom shell. That way even the smallest diatoms can bring themselves, or at least their offspring, back to full size.

Sexually reproducing protists—giants in a world of bacteria—evolved into a new variety of complex patterns tailored to different lifestyles and environments. The co-evolution of monera and protists with these environments led to one improvisation after another, including the protists' formation of a new kind of cooperative that led to multicelled creatures, and thus to the remaining three kingdoms of life—fungi, plants, and animals—in all their visibly fantastic variety.

O   O   O

We have watched the Earth transform itself from a fiery ball to a crusty planet whose skin came ever more alive as giant molecules and enzymes formed, then packaged themselves as bacteria. We've seen living masses of microbes transform themselves as they discovered new lifestyles, rearranging and recycling minerals, creating the atmosphere, and so on. We saw how the monera collected themselves into the much larger protists, inventing nuclei and stumbling on sexual reproduction. Now our story continues with the evolution of ever larger and more complicated creatures.

Some protists began living together in colonies by sticking together after division rather than floating off on their own, though each protist in the colony remained as independent as if it had gone its own way. But just as protists evolved when various monera began working together as cooperatives, protists living together as colonies eventually took the next step of communicating with one another and developing joint plans.

After all, communication had always played a major role in Gaian life, from the bacterial Web to the DNA-RNA-protein communication systems between the nucleus and cytoplasm of eukaryote cells, and the communications systems between cell membranes or walls and the outside world. Now it was time to develop communications among eukaryotes, and one way in which they did it was by sending chemical messages to each other.

This chemical communication made it possible for the individual protist cells to harmonize the things they did—such as beating their cilia oars together in rhythms that moved the whole colony smoothly along in one direction. The ability to communicate soon became useful in many new ways of cooperating, especially in divisions of labor among different cells in protist colonies, thus beginning the evolution of multicelled creatures.

A most peculiar and fascinating in-between step in the evolution of multicelled creatures from single cells is the appearance of slime molds. Scientists have devoted whole careers to studying these strange creatures that seem to be part protist, part fungus, part animal. Slime molds start out as separate amoeba-like creatures, but when their food source runs out, they emit chemical messages that attract them to one another until they gather together into a visible slug-like community. You can see their jellylike mass sometimes on the underside of a rotting log or leaf. Some varieties form a large sheet of jelly.

The slug-like community can actually move itself about like a brainless worm, but eventually it stops and begins sprouting stalks that form fruiting bodies on their tips. These then release spores—tiny dried-up packets of DNA and other cell materials—the way any self-respecting mold would. The spores blow through the air and, after settling in a new moist place, form new amoeba-like creatures to start the cycle all over again.

Slime molds thus are capable of specialization and cooperation under hunger conditions, if not otherwise. Note that we have now found hunger as the prod behind the cooperative evolution of nucleated cells, the invention of cooperative sexual reproduction, and the evolution of multicelled-creature cooperatives—all creative responses very different from the competitive struggle Darwin attributed to food shortages.

In other ancient protist colonies that did not lead the double lives of slime molds, some of the cells became specialists at making food, others at catching it, still others at breaking down and digesting food. In some colonies there were specialized cells to move the whole thing about; others contained specialized cells for sticking it tight onto rocks. The first multicelled creatures were cooperative colonies of protists, just as the first protists had been cooperative teams of monera— cells. Our present human process of globalizing seems to be forming us into a new planet-sized multi-creatured cell, in what we might call a fractal biology of repeating evolution-ary patterns.

Let us go on with the story of the first multi-celled creatures for now.

In some colonial creatures, certain cells began specializing in the sexual reproduction of the colony as a whole. That meant a few cells could reproduce the whole colony, instead of each cell in the colony reproducing itself. This was a big step in the evolution of colonies into creatures and in the evolution of embryos as a way of starting new generations. For the continued life of the species, all the other cells now depended on the specialized cells that produced gametes—those haploid chromosome packages we saw diatoms making. We call them gametes because they were not yet either the specialized eggs and pollen of plants or the eggs and sperm of animals. Early gametes from different parent looked alike, though they had to pair to make a new being.

The development of multicelled creatures that reduce them-selves back to single-celled creatures in each generation to carry out their reproduction, brought inevitable death by aging into Gaia's dance—a new way to ensure recycling. Bacterial progenitors, remember, do not die except by unfortunate accidents, such as being burned by ultraviolet. These in fact happen often enough to keep bacterial populations within bounds, but they do not die of old age. Instead, they phase out their physical and genetic identities over generations of gene-trading offspring.

Multicelled plants and animals, however, leave their bodies behind as their genes continue on in new generations. DNA is the oldest living survivor in all nature. From a microbial point of view, the large multicelled bodies cloned from eggs and seeds have no further value as new generations emerge, except as excellent sources of food.

The fact that death is necessary for multicelled life to continue virtually without end has been hard for us humans to grasp and accept. If new creatures kept coming to life without others giving up their lives, the supplies in the Earth's crust would soon be used up and the mass of creatures would all die together of crowding and starvation, as we humans are rapidly learning from our successful efforts to increase food supplies and delay death.

With the death of creatures so that others can recycle the materials of their bodies, life can go on and on. In fact, the way living things die to make way for new life in Gaia's dance is very much the way things happen in any dance. Every dancer knows that each dancer can only perform one step at a time—that old steps must be abandoned so that the dancer's body will be free to perform new ones, which may then repeat or change the pattern of old steps.

Gaia, our living Earth, has lived for billions of years and has billions more years of life ahead. Our individual bodies will die and be replaced by others, much as the cells of our own bodies are constantly dying and being replaced by new cells. Every seven years or so we are in fact a wholly new person through such replacement of cells, yet we only see the changes in our bodies as aging, not as endless newness. In the same way, from Gaia's point of view, there is no death—just endless replacement of old cells in her body with new ones.

Every atom that is now, ever was, or ever will be a part of us will live on somewhere in Gaia's ever-evolving dance for billions of Earth years yet to come. Even after the Earth dies, those atoms will live on as part of our galactic dance, some perhaps finding their way into new living bodies of new planets.

O   O   O

Animals, as part of our own planet, were a marvelous evolutionary development in the face of yet another problem. Some protists, after running out of food, were unable to make their own food from light because they contained no bluegreen bacteria, or their chloroplast descendants. We are not sure whether our own protist forebears never took in any bluegreens or whether they took some in and later lost them. We do know that plants have always had both chloroplasts and mitochondria, which allow them to make food using sunlight and to burn food using oxygen. Animals, including ourselves, can only burn ready-made food.

This means that plants could—and still do—live their whole lives sitting in one place, making their own food, while animals had to evolve ways of going after their food. Animals, as we will see, evolved all sorts of equipment, from eyes and ears to feet and wings, to heating and cooling systems, to nervous systems with brains for organizing all this complexity, just to help them chase after food—and all because they had no chloroplasts!

Nature is, of course, never quite as orderly as we would have her, so she managed to leave around some puzzles such as the giant green clam, an animal that does have chloroplasts and uses them to make emergency energy from sunlight, though it is clearly an animal in every other way.

Among the earliest multicelled animals to evolve from protist colonies were polyps. Luckily, there are still many living polyp species that match ancient fossils and so give us clues to their early evolution. Actually, polyps look more like plants than the animals they are. Sea anemones, which look like flowers, are polyps; forests of coral are huge polyp colonies.

The polyp animal is shaped like a tube with a flowerlike circle of tentacles at one end around its mouth. The other end of the tube is stuck to a rock or to the body of another polyp in its colony. And there it stays. It is a simple animal with a body organized to catch its prey in its tentacles and stuff the food into its mouth.

Still, many polyps have rather amazingly complicated cells along their tentacles. These cells have a special name; we call them *nematocysts*—meaning 'thread bags'— because they evolved from ingrown cilia that grew into extremely long, thin, hollow threads and became very specialized in their job. When prey touches one of these surface cells, the long coiled thread shoots out under the pressure of liquid filling it, tangling the victim and paralyzing it with poison barbs. Nematocysts are a wonderful example of the amazing patterns of organization that nature has worked out even within the cells of the smallest and simplest of creatures. Nematocysts are such good self-contained weapons that other creatures, after eating polyps, may not digest the nematocysts but may, instead, keep them for their own use in catching prey.

Polyps reproduce by budding like bacterial forbears. This job is sometimes assigned to certain members of a polyp colony, which are fed by the others so they can concentrate on their important work. In some species the polyp buds grow up stuck to the parent, but in others something much more interesting happens. The newly budded polyp breaks off, flips over so its tentacles hang down, and floats off into the sea. As it grows, it becomes a glassy bell or umbrella with a softly fringed edge of trailing streamers—a jellyfish, as we call it, though it is not a fish at all. Its proper name is *medusa*—a name taken from the ancient myth of a woman who had snakes on her head instead of hair.

Medusae are a much more adventurous stage of polyp life that learned to reproduce sexually. Some species tried having both sexes in the same individual—as flowers and earthworms have them—

while other species began making separate males and females. In any case, all medusae produce female eggs and male sperm, which fuse to make baby medusae. The baby medusa is so different from its parents that it, too, gets its own name. We call it a *planula*. The planula is a long, flattish blob that rows itself about freely for a while using a fringe of cilia. Then it settles onto a rock and sticks itself tight to grow into—a polyp.

The life of a polyp is thus a matter of *metamorphosis*—changing form from planula to polyp to medusa. Such metamorphosis was later repeated in evolution, in butterflies and moths, for example, like so many other earlier step patterns that are woven again and again into the later dance. Polyps in countless variety still abound in the seas, looking much as did their ancient forebears. Yet sometime, somewhere in the dim past, some of them became discontented with this three-stage metamorphosis, which always came back to a sedentary phase, and went on to invent more adventurous lives.

9

From Polyps to Possums

One of the many remarkable things about life is its memory. The process of life creates and stores information—not just the kind of information needed to reproduce each new body from a single tiny cell, but remembered information about much of its evolution over eons of time. The way each of us came into being shows us something of the whole dance of evolution since the time of the first protists. Before we are even born—in just a short nine months—we repeat many steps in a billion years or so of our evolution as Earth creatures.

Each of us began as a single cell, much like an ancient protist that began its life by sexual reproduction, as the offspring of two parents. This new-creature cell divided again and again, cloning itself first into two cells, then into four, eight, sixteen, thirty-two, sixty-four cells, and so on until there was a simple ball-like creature very much like a protist colony. This creature in its early stages lived like other protist colonies in a salty sea—a special uterine bag of salty liquid duplicating our real ancestral sea.

Our own embryonic colony of cells soon specialized and turned us into multicelled creatures. If you could watch a movie of your own development—you may have seen films of human embryos developing—you would see the ball-colony change shape as its specialized cells divide again and again. One side of the ball dents in to form a groove where the backbone will grow. A lumpy head appears at one end. Soon it looks like a tiny fish with gill slits; dark eyes bulging in its big pale head, and a tail that starts twitching. A little later it looks so much like the developing embryos of frogs and turtles, then chickens and pigs, that it is hard to say what it is going to become. Even when our arms and legs have budded from our bodies, we still look much like other animal embryos.

Slowly we continue our formation as we roll about comfortably in our warm sea. Our tails shrink to nothing, our brains grow bigger, our arms and legs and faces become human. Finally, after nine months, we leave our maternal sea to begin breathing air.

How fascinating that this memory of the Gaian life dance is relived by each of us, reminding us of just who we are, where we came from, how we are related to all other species and to the whole dance of life—the evolutionary dance we traced at the end of the last chapter to our discontented polyp ancestors. Let's continue now to see what made them evolve into more complicated animals and how *they* came to leave their ancestral sea.

O   O   O

Dull as the polyp's life cycle was at the beginning and end, the middle stages as medusa and planula permitted some adventure, some spreading into new environments. Somewhere along the line, certain young planulae seem to have felt their difference and rebelled against settling down and growing into a dull adult life as polyps. They began growing up straight into medusae, simply skipping the whole attached polyp stage. Others began sexually reproducing themselves as planulae, skipping even the medusa stage and evolving, generation by generation, into various species of wormlike and then fishlike creatures.

The evolution of new species from the baby stage of a parent species is known by evolutionists as *neoteny*—meaning 'stretched youth.' Neoteny is a kind of evolutionary step backwards in the dance when a species finds itself at some adult dead end or blind alley, when for some reason the body organization of the grown-up stage evolves to limits that get it stuck.

It seems that when a species becomes highly specialized for a particular and successful way of life, it loses variety in its DNA and ends up with a very fixed body structure that can no longer change or evolve. We can see such dead ends in polyps living today, or in sharks, which are still very like their ancient ancestors, having evolved no further. But in some ancient species, as we said, nature took advantage of the fact that baby planulae, which did not yet have the specialized bodies of adults, could still change. We can trace the descent of some free-swimming creatures back to ancestors who were very likely unspecialized planula babies. Later we will see that neoteny produced other interesting new species, including our own!

Among the new parts of the more complex bodies evolving from planula-like ancestors were networks of nerves. These seem to have evolved from in-turned rowing cilia that became tubes for sending messages from cell to cell. Once animals evolved nervous systems they kept improving them and never gave them up, for these communications systems made it possible to organize ever greater numbers of cells within a single creature, just as nuclei had made it possible to organize ever greater numbers of cell parts within a protist.

Organizing now meant actually forming organs—grouping cells into body parts specialized for particular jobs, such as guts for digesting food, hearts and blood vessels for circulating supplies, eye spots for telling light from dark and later making images that helped identify food or predators. Remember that animals were pushed to evolve complex organs because they had to hunt for their food rather than make it themselves. The other side of that coin is evolving ways to avoid being eaten yourself.

An extra tube, stiff but bendable, evolved along the main nerves running from one end of early wormlike animals to the other. This protected the delicate nerves against damage. Later the tube wrapped itself right around them to become a backbone. But long before that happened, the tube was useful in another way. Long cells, good at stretching and shrinking, attached themselves to it, pulling the tube into wavy patterns. That was the beginning of muscles, which gave creatures an important new way of moving themselves—from the inside, rather than by outside rowing hairs or tails.

Clumps of nerves at the head end of creatures evolved into simple brains; the eyespots evolved into true eyes with lenses to focus light and retinas to make images of what went on in the environment. Squid and cuttlefish have bodies that remind us of tube-shaped polyps with tentacles, but they have evolved eyes not unlike our own. Lyall Watson pointed out the biological mystery of why such wonderful eyes evolved in creatures with so little brain to understand what their eyes can see.

The muscles of squid and cuttlefish evolved into another way of travel—by jet! Their tube bodies take in water and the muscles squeeze it out suddenly, shooting the creature along. They, along with their octopus cousins, also evolved a way of hiding even out in the open, by making supplies of black ink to squirt like a cloud into the sea around them. Octopuses also evolved bigger brains, and are smarter than one would guess.

In some species of early animals, mouths gave up tentacles to grow a single big sucker, as in some eels, or to build jaws and then teeth, which became very popular. Different species experimented with inner bones and outer shells to hold their bodies together and protect them.

So, one evolutionary invention led to another, sometimes improving an old pattern a little at a time, other times by actively reorganizing available genes quite quickly to produce a new pattern. Every time a creature's DNA replicated itself in the process of forming egg or sperm cells, and every time these came together with gametes of the opposite sex, if not at other times as well, genetic information could be shuffled. Other sources of genetic change now seems to be through the activity of DNA itself, in direct response to situations, and through gene-trading bacteria moving in and out of the multicelled creatures' cells. The big mystery is just how all these processes are coordinated within and among species. Our best hope of solving it may lay in the physics of cosmic consciousness, continual creation and non-locality, as they increasingly understand fundamental levels of non-physical communications.

Looking into the past, using fossils, microscopes and other technologies, shows us that most genes existing now were developed in very ancient times, just as we already saw that bacteria had developed almost all the molecules and proteins that now exist. The new DNA development since early multicelled creatures seems to lie in the *regulator genes*, which organize simple genes into more complicated genetic patterns. This makes it possible to keep evolving endlessly new patterns out of the same basic genes

Some of the most fascinating of today's biological puzzles are arising in genetic engineering—a field in which we take advantage of the microbial ability to transfer genes and get them to do it for us, though the results are not as predictable as desired. We implant genes and sometimes the microbes themselves into species we wish to alter—to make them 'better' in texture, flavor, nutrition or to resist herbicides we want to spray on their fields to kill weeds, and so on. For example, microbes emitting particular toxins are implanted into corn seed; when the corn grows up every cell contains these microbes and emits the toxins that kill insects. The puzzles arise when the implanted species remove the genes we implant, or transfer them to weeds, making them equally resistant to our poisons. We would do well to study nature's practiced ways before we leap into our own attempts to improve on them.

Much about the organization and reorganization of DNA may be better understood when we understand more about the brain. The DNA pool of bacteria, the cellular nucleus, and the brain are all natural systems for receiving, storing, and processing information required by organisms. The efficiency of the Gaian way of life in using the same schemes over and over in ever more complex arrangements suggests that these three systems are likely to have a good deal in common. We have learned that the brain is more coordinator than dictator of the body's physiology and behavior—a kind of central clearinghouse and resource center for the body as a whole. Perhaps the organized DNA of cell nuclei is similar, not dictating what the cell does, so much as being used as a resource center by the cell as a whole.

O   O   O

A step at a time, over many millions of years—though some steps as we know were faster than others—the seas filled with an incredible variety of living things co-evolving as ecosystems. Or, we might say, from our Gaian viewpoint, the seas with their ever-renewed supplies of salts and minerals washed from land rock turned themselves into a living soup of plankton and larger creatures.

Many species continued the 'plant way of life,' though they were not yet true plants, evolving seaweed colonies from simple algae. Some had parts that looked like roots, stems, and leaves, though true roots, stems, and leaves evolved only much later on land. These chloroplast-containing creatures took over bacterial ways of fixing nitrogen, took other building materials from the sea and from the sea bottom, which was rich with decaying bodies, and continued making food using solar energy. Their usable nitrogen and the carbon was taken from carbon dioxide and built into their bodies, then passed on to animals who ate the algae to build *their* bodies.

Some of the carbon was recycled, but much of it was buried in sea floor sediments of dead protists, algae, and animals. Over billions of years of evolution this process, plus a similar carbon burial on land after plants evolved there, used up most of the early atmospheric carbon dioxide. It and other early atmosphere gases were gradually replaced by a balance of mixed gases that were almost entirely the products of living creatures and were just right for their continued survival and evolution—gases constantly burning each other up, constantly being recycled and replenished by living creatures.

Bacteria and protists continued to live among the larger creatures in far vaster numbers, working at the rebuilding and balancing of the atmosphere and the chemistry of the seas, as well as providing the larger creatures with food. Recall that in each ecosystem, the member species co-evolve as they affect each other's lifestyles and forms.

Over time ecosystems mature from a few species that may compete in a juvenile way for food and space to a mature ecosystem in which many species balance their lives cooperatively.

Plants did not become as complex as animals because nothing pushed them—life was too easy. They had no need to go after their food, to see it, to grab it, to digest whole other creatures. Animals, who did have to do these things, tried out many ways of building themselves. Some evolved hard coverings like those of clams, snails, barnacles, spiny starfish, and sea urchins. Some, like crabs and lobsters, successfully tried out legs in pairs. Others, like worms, squid, and octopus, stayed soft. Still other free-swimming forms became sharks and the bony ancestors of modern fish. But no matter how different their shapes, they all evolved muscles to move with, blood to circulate supplies, eyes to see with, and nervous systems with brains to coordinate their ever more complicated bodies.

Together, this great variety of living things created different ecosystems for themselves and for one another on the sea's floor, on its surface, and in the shallows near shore. The first such systems with large creatures appeared so suddenly we refer to them as the Cambrian Explosion. This is what we identify as the start of the Paleozoic Era—meaning 'ancient animal period'—around half a billion years ago. Our best fossil find of this explosion of creativity is the Burgess Shale in western Canada, with a wide variety of creatures from very large soft-bodied quilted creatures to smaller armored creatures such as *Opabinia*, with a long, flexible but fanged vacuum cleaner hose of a mouth. Some look like moonwalkers or weird lifeforms we imagine on other planets. All together they look completely unlike any ecosystem of today, and indeed they went extinct long ago.

O   O   O

The stardust that formed our rocky Earth had come a long way in rearranging itself. But while countless small bits of the planet's crust were turning into living creatures, the crust as a whole had broken into pieces that slid about on the softer molten insides, like the armor plates of an armadillo. Ever new eruptions of lava pushed the plates apart by adding cooling rock to their edges, while their opposite sides slid under the edges of other plates to make room.

We introduced them in Chapter 2 as tectonic plates, their name coming from the Greek word for 'builder.' And, indeed they built the shape of the world we know. The thickest parts of tectonic plates are the landmasses we call continents that stick up out of the seas. Over the past three billion years these thicker continents have repeatedly moved together and apart, with half their landmass submerged when they are most spread out. During the spread out phases, such as the one we are in now, Gaia's temperature drops an average of ten degrees centigrade, causing waves of recurring ice ages in cycles of about a hundred thousand years.

At the time the dinosaurs evolved, hundreds of millions of years after the Burgess Shale creatures, all the thick parts had moved together into one huge supercontinent called *Pangaea*, which means 'all Gaia'—a name chosen well before Lovelock's Gaia hypothesis. Ever since then, this landmass has been breaking up to form continents separated by oceans and seas. The Atlantic Ocean is still getting wider year by year as South America and Africa are pushed ever farther apart.

Before they became the Pangaea supercontinent, the continents had been separate, but they had then dominated what we know as the southern half of the Earth. As Pangaea formed and split up again, the continents moved northward and apart from one another. If you could see their movement over billions of years as a film, you would see them riding the slowly swirling soft insides of the Earth.

O   O   O

Let's go back to pre-Pangaean times now, to the next big step in the evolving dance of life—the great land adventure, when some creatures took to dry land and continued to evolve there.

Of the organisms that were larger than microbes, algae and funguses got to dry land first, paving the way for animals by multiplying into rich food supplies, especially by evolving into plants—just as bacteria had paved the way for plants by breaking up rock and reproducing themselves, thus starting a supply of soil.

Perhaps the migration of creatures onto land began when the algae were left high and dry for hours every day by Moon-pulled tides. To survive, they adjusted their bodies, learning to live both in and out of water. Algae and funguses first appeared on shores, then their spores were blown further inland by wind. Wherever bacteria had made enough soil and there was enough moisture from dew or rain, such spores developed.

Fungi are a whole kingdom in themselves, including molds, yeasts and mushrooms. They dissolve rock for food by excreting acids, and digest organic food before consuming it by excreting enzymes onto it.

The first plants, evolving yet another new kingdom, were mosses growing close to the ground along with cooperating teams of algae and funguses called lichens, which look like very close-cropped plants and come in a variety of colors. It took a long while for plants to develop strong roots and taller, stiffer bodies that could evolve into ferns and trees. The major new structures plants had to develop to support their bodies in air and carry water from roots to all parts were *vascular* systems—stiff tubes, or veins, running the length of plants. The whole plant kingdom is divided into vascular and non-vascular plants.

Animals followed more slowly onto land. *Arthropods*, the first to come ashore, are jointed-foot creatures with their hard skeletons outside— from the Greek, *arthro* meaning 'joint' and *'pod'* meaning 'foot.' Arthropods evolved in the sea—the Burgess Shale's Opabimia was an example—but they also learned to breathe air in coming to live on land, where some of them evolved into light-weight insects, while their relatives in the buoyant sea evolved larger bodies, such as those of crabs and lobsters. Horseshoe crabs, like sharks, are like those bicycles in a jet age—Cambrian creatures still going strong today!

It was just as fungi, plants and insects were getting their real grip on land that the first great extinction occurred—around 440 million years ago—when one of Gaia's great waves of ice ages formed great land glaciers and chilled the seas. More than half of all her life forms died out—the least affected being the adaptable bacteria and pro-tists—and it took her 25 million years to recover her biodiversity.

After the extinction, a new type of animal emerged from the sea. These animals, like the early shore plants, evolved ways of living part-time in the sea and part-time out of the sea. We call them amphib-ians—amphi meaning 'double' and bios meaning 'way of life'—creatures who live double lives, on the land and in the sea. While arthropods have their bony skeletons outside their bodies, amphib-ians were fishlike creatures that crawled on fins and developed lungs to breathe air. Eventually they transformed their fins into short legs, and later we will see that some species evolved into reptiles. Most of the ancient species of amphibians died out as other animals evolved. Among those still living today are frogs and salamanders.

Only a hundred million years after the first mass extinction there was a second, again related to cooling climate change, again doing in half the species of Earth. The third mass extinction happened only 37 million years later, so it took a hundred million years altogether for full recovery. During that recovery plants had their time in the Sun, as we say—great carboniferous forests with giant tree ferns, ginkos and cycads growing up and thriving for 70 millions of years, generation by generation, removing carbon from the atmosphere and burying it with their bodies as they were pressed underground to become coal and oil. It was during this era that Pangaea the supercontinent was assembling herself from older pieces named Laurasia (which would later again break off to form Asia, Europe and North America) and Gondwana (which later broke off to form South America, Africa, Australia and Antarctica).

Insects thrived in the great forests; amphibians moved inland, inventing self-contained eggs and splitting into two lineages: the *synapsids*—meaning 'with arch' (in their bony skulls)—that evolved into large four-footed *tetrapods* that evolved in turn into mammals and humans, and the second lineage: the *reptiles* that evolved into turtles, snakes, lizards, and *archosauromorphs*—meaning 'ruling lizard forms.' (Note: *archeo* means ancient, *archo* means ruling.) Guess which became our favorite dinosaurs and pterydactyls?

In that great age when the ruling reptiles stalked the Earth as the largest of its creatures for 160 million years—two to three times as long as mammals have been around, some forty times as long as humans have existed. The fossil record shows many shapes and sizes of reptiles and lets us know what important roles they played in evolution. But a quarter billion years ago, just as the first lizard began gliding through the air near the end of the Paleozoic era and before our favorite huge beasts appeared, 95 percent of all existing species disappeared in the fourth and greatest extinction. About the best thing we can say about it is that its recovery period—the *Mezozoic*, or 'mid-animal-era'—brought the world an explosion of color in flowers along with *triceratops, brontosaurus* and early birds.

Flowers, those marvelous sex organs with their wonderful blaze of color, brought a special kind of beauty, but that beauty was practical as well. With flowers, plants achieved their full two-parent sexuality, but with both sexes housed in the same individual. Sitting still in one spot, they needed a way to spread their genes around to other potential mates they could not reach themselves. Flowers gave plants a way of attracting birds and insects to cooperate with them in getting the male pollen of one plant to the eggs of another.

It is interesting to note that 60 percent of all known species are insects, while less than *one* percent includes all birds and mammals! One third of the insect species—one fifth of all species—are beetles. It would be easy to argue that beetles, with their four-winged armored bodies are the most successful of Gaian creatures.

Birds, we can see clearly, descended from the still half-reptile *archaeopteryx*—'ancient wings'—and its cousin the *pterodactyl*—'feather fingers' (both the words 'wing' and 'feather' come from the Greek root ptery). Their fossils, as we said earlier, show us leg bones evolving into wing bones, jawbones into beaks, scales into feathers. These early birds were far larger than today's, with wingspans up to twelve meters. People build flying models of them to see how such great beasts could stay in the air.

The great beasts of the Dinosaur Age grew up to twenty-seven meters long and some species towered very tall on their hind legs. Scientists still argue about just what dinosaurs were, and at present it looks as though they were rather odd creatures—neither reptile, nor mammal, nor bird, but with elements of all three. During their reign, the land world buzzed with flying and crawling insects and grew ever greener with plants and their colorful flowers. In the sea world, lovely protists were filling the oceans to recycle minerals, while great marine lizards called *mosasaurs* and *icthyosaurs*—'fish lizards'—swam over bottom-dwelling *ammonites* and clams and other creatures.

Late in the age of dinosaurs, *therapsids* were evolving into true mammals with a kind of croco-dog-bear look, just as *archaeopteryx* and *pterodactyl* were transforming their lineages into birds. Warm-blooded mammals, who keep their own body temperature constantly warm, as dinosaurs seem to have pioneered, and keep their babies inside their bodies, rather than in eggs, until they are ready to be born, were the last kinds of animals to evolve. Though mammals and birds were until not long ago thought to be the only warm-blooded creatures, we now know that some fish, as well as some dinosaurs, also evolved this feature. Red-blooded tuna, for example, keep much warmer than the surrounding sea, with systems that are extremely efficient in preventing heat loss.

O   O   O

The really big boost to mammalian evolution was given by the same catastrophic event that spelled utter disaster for dinosaurs. About 65 million years, the Mezozoic era ended abruptly as a huge meteor plunged into what we now call the Caribbean, first incinerating life, then freezing it in cold temperatures as a black cloud of debris spread around the Earth, causing the fifth great extinction—the last until we humans initiated the present sixth one!

The great dinosaurs disappear, as do archeopteryx and pterodactyl, from the fossil record after this great catastrophe. The weather, the climate, the whole environment on which they depended was gone and the huge specialized creatures who had ruled Gaia for so long could not change fast enough to survive such sudden change. But just as new types of bacteria took over when many of Gaia's early bacteria had died of oxygen poisoning, new kinds of animals evolved now from the small ones left after this great extinction. Gaia has shown again and again the ability to recover from disasters, always continuing the dance of life in creative new ways.

So life continued in the new age, the Cenozoic, with particular species of plants and animals evolving particular bodies and ways of life to balance and harmonize with one another as parts of the great Gaian system in which they evolved—a system that went on working as a single being to regulate the Earth's temperature, chemistry, and weather.

Living things may even help tectonic plates to move by weighing them down as their bodies turn back to rock, or at least by providing chalky layers that help some of the weighted plates slide under the edges of others. Pangaea was breaking up, splitting Africa from South America and separating Greenland out between America and Europe, the Atlantic Ocean flowing between all these pieces. India, at that time, had moved about halfway from Antarctica, where it began, toward Asia, where it was to get stuck in place, pushing up the Himalayas as it crunched into the greater landmass.

Somewhere around 50 million years ago, just before India struck Asia, it pushed the seafloor up, forming a warm shallow sea called *Tethys* that teemed with plankton life and lured some early wolf-like creatures back to the aquatic life. This is the apparent origin of the sea mammals we call *cetaceans*, which include whales and dolphins. First the wolf-like creature reverted to a sort of hairy crocodile amphibian stage, then they took seriously to the seas, giving up fur and feet for smooth skins and flippers. *Pinnipeds*— 'fin-feet'— also became aquatic mammals, such as sea lions and walruses.

The Earth was beginning to look as it does today, though the continents were still closer together than they are now. As land bridges between them disappeared under water, plant and animal species were stranded on separate continents to continue separate paths of evolution. This is why many species alive today, like the kangaroos and koalas of Australia, are found only on a particular continent. Some species of different continents, such as the alligators of North America, the caimans of South America, and the crocodiles of Africa, are still recognizable descendants of the same Pangaean ancestors despite differences due to their later evolution.

Remaining reptiles in the Cenozoic were, as they still are, cold-blooded animals, their body temperature rising and falling with the temperature of their environment. Like the birds descended from them, we saw that reptiles lay eggs. But unlike warm-blooded birds, they rarely take care of their young. Their brains are so simple that most reptiles can't even recognize their own babies when they hatch. In fact, they have been known to eat their own babies—not because they are vicious or cruel, but because they cannot seem to tell the babies from other edible things. This challenge to reptile babies made them evolve into creatures that run very fast very early. A baby reptile, in fact, is almost as good as a grown-up at doing most everything reptiles do.

Reptile behavior emerges directly from reptile genes, bodies, and physiology, without benefit of very much thought or choice. They do pretty much the same things in the same old ways, not having enough brain to think about or change them. They hunt food, show off to win their mates, huff and puff at strangers who come into their home territory, and fight if the stranger is not frightened away. That's about all. They don't even sleep, but just settle down quietly when night cools them off.

Early mammals were quite different, with their lively warm-blooded bodies, the female ones keeping their offspring inside until time for birth and feeding them on mothers' milk from the mammary organs, which gave them their name. Many early mammals were active by night, having evolved a pattern of waiting till dinosaur types went to sleep before going about their business. These included small tree-dwelling primates with stereoscopic depth vision adapted to seeing in the dark and nimble fingers to pluck their food from branches. They carried their young on their backs as they swung nimbly through the trees.

True mammals branched into two types. One branch, the *marsupials*, which includes kangaroos and opossums, gives birth to very undeveloped babies that stay in pouches on the outside of the mother's body after they are born, until they are ready to live on their own. The other branch grows its babies to a later stage deep inside the body until they are born.

Some of today's small mammals show links back to the early mammals that evolved from reptiles. The platypus, for instance, is a strange, furry, warm-blooded aquatic animal that lays eggs and has a duck-like beak and webbed feet, as though it couldn't decide whether to evolve into a bird or a mammal. A few mammals later *de*volved their legs back to flippers and fins when they returned to the sea to become seals, sea lions, manatees, whales, and dolphins.

Possums are one of the most primitive, or antique, species of mammal living today. They seem to have changed little since they evolved among the dinosaurs, so they are important in the study of evolution. Among other things, they may have been the first animals to sleep and to dream.

10

From Possums to People

We are still not sure why warm-blooded animals dream during sleep. Sleeping and dreaming are special behaviors that evolved in mammals and were passed on to their descendants, yet no one really knows why. Scientists used to think sleep restored worn bodies, but there is no good evidence for this theory. Some people sleep very little, even never, and have no problems as a result. For most of us, however, sleep is clearly an unavoidable aspect of our physiologies. That makes it highly likely that sleeping and dreaming evolved because somehow they did our ancestors some good.

One guess is that sleeping was a way to keep warm-blooded bodies, which worked just as well by night as by day, quiet and in hiding during the hours when dinosaurs hunted, as mentioned above. Perhaps dream images of dinosaurs kept the mammals just close enough to waking so they could get up and run fast if a dinosaur poked its head into the nest. This theory seems to fit mammals that are active by night. But shrews, small mouse-like mammals, are active day *and* night, while bats, which are small mouse-like mammals with big wings, sleep almost the whole day and night. Many mammal species, including humans, do their sleeping at night instead of by day.

Another theory says that we are put to sleep by a chemical that daylight builds up in our bodies, making us more and more tired by nightfall. That still does not explain what good sleep and dreaming did us. All we know for sure is that they were among the many new kinds of behavior organized in the larger, more complicated brains of early mammals such as the possums. Perhaps we sleep simply to prolong our lives, for sleep does slow down body activity and thus makes a body last longer. Bats live about five times as long as their super-active shrew cousins with the same body size, though both species have about the same number of heartbeats per lifetime! But long life in individuals does not necessarily mean their *species* survives longer, as it is clear that bat and shrew species have survived equally well.

Possums, besides being warm-blooded, hunting by night, and sleeping with dreams by day, take care of their young. Caring for babies through an extended nursing period became the basis for more complex social behavior. Bigger brains had room for more complicated ways of seeing, feeling, and doing, of understanding and acting. Among the new ways of seeing and feeling was the ability of parent animals to recognize their babies and feel the need to care for them—to feed them, clean them, protect them, and teach them, as both birds and mammals do.

Again we must note that nature does not keep things in neat human categories, and so we find insects, such as wasps, providing for their young in advance of their hatching by storing food, and frogs that keep live young in their stomachs, regurgitating them now and then to see if they are ready to function on their own.

Some species of fish have evolved quite elaborate care of live young by their parents, though they are not warm-blooded. Seahorse daddies keep babies in kangaroo-like pockets, and some daddy cichlid fishes keep hatched babies in their mouths. After they learn to swim on their own, the fathers herd them and let them back into their mouths any time danger threatens.

Still, if you could watch and compare a fish father and a cat mother caring for their young, you would quickly see that the fish has simpler movements and acts more automatically, while the cat often seems to have a choice about what to do next. The tendency toward particular patterns of behavior evolved in animals together with the structure of their bodies. Such innate, or built-in, behaviors—say, nest building or courting rituals—are popularly called 'instinctive,' though most scientists, having dropped the old concept of instincts, call them 'species-specific behaviors' or simply 'innate behaviors.' Whatever we call them, the animal performs them without having learned them in the way we understand learning, and so they appear automatic, performed without choice or flexibility. Sometimes they are therefore called 'fixed action patterns.'

As evolution transforms creatures from simple animals to amphibians, to reptiles, and then to mammals, we find nervous systems getting more complex and behavior becoming more flexible. As brains grow larger and more complicated, some of the innate behaviors loosen up, giving the animal more freedom of choice in responding to its environment and to its own inner urges. The brain is increasingly a flexible coordinating system for changing behavior to fit changing circumstances.

As animals gained more freedom in what to do, and how and when to do it, they gained more freedom in acting on feelings as well as information and in learning new behaviors. Mammals clearly began to show what we recognize as feelings, and some of these feelings seem to be the beginning of evident love in animal evolution. Animal mothers apparently felt good staying close to their babies, feeding them on milk from their own bodies, licking them clean, and hiding them from danger. Whenever mothers and babies got separated, both felt distress and cried. They got back together by listening to each other and tracking the cries. That was the beginning of the voice communications we humans have evolved into complex languages that can be used to express extremely complex ideas and information. Birdsong is a more innate pattern of voice communication, though baby birds often have to learn the exact pattern for their species from the adults. Whales and dolphins have reasonably complex languages they teach their young.

One of the most important things mammals learned and taught one another was to play. Play in baby animals seems to be practice for more serious grown-up behavior such as hunting or winning a mate. You can see wide varieties of baby mammal play, from kid goats butting one another to puppies hiding from and pouncing on one another. Mother cats clearly show their kittens how to practice hunting on one another without being too rough, and the kittens come to enjoy the activity. Ever since it evolved, play has been an important part of social life—the life of animals living together, making a living together, communicating, and caring for one another.

Mammals such as cats, dogs, and monkeys have flexible bodies that make it easy for them to care for babies and to tumble about in play. Flexible feet with toes and nails, or claws, can become very useful paws with which an animal can do many things, as we see watching raccoons manipulate food in complex ways. Stiff-legged mammals without flexible toes, such as goats, antelope, and horses, are specialized for climbing and running, but not for tussling or washing their food! Thus they lead simpler lives, grazing rather than hunting other animals, running from their predators, shielding their babies as best they can. While their strong, slender legs and hooves make them sure-footed runners, such legs are too specialized for baby care and play, so their young stand and run on their own stiff legs early.

Mammals of all shapes and sizes—some very specialized, some more flexible—branched off from early possum-like ancestors. Each new species created new steps in the Gaian dance of life as it wove itself into a complex environment. While the oceans filled with animals from microbial plankton to sea snails, from fishes to mammalian whales and sea lions, forests grew thick with colorful insects, birds; and furry mammals, plains rang with the clattering of fleet-footed hooves or shook with the heavy tread of pachyderms.

With the dinosaurs gone, mammals were the largest of land animals, just as trees were the largest plants. All through evolution, larger and larger individuals had evolved in both the plant world and the animal world. But there was a size limit beyond which individuals did not work so well. Trees that grew too tall could not stand upright in storms or pump water to their highest leaves. Giant dinosaurs were too large to survive a catastrophe that smaller creatures did survive. Mammals the size of whales and elephants seem to be about as big as the Gaian life system can manage successfully. Larger bodies, among other problems, would have trouble getting enough food and oxygen. An animal ten times as big as another in each direction weighs a thousand times as much and needs a thousand times as much food. The large animals that have evolved are very few in number compared with smaller species, down to the most numerous of all, the endlessly hardworking bacteria.

While some Gaian animals reached size limits, there seemed to be no limits to variety in their body designs and behaviors. They evolved countless wonderful ways of swimming, slithering, crawling, running, climbing, and flying; of hunting, fighting, playing, and learning. They had senses to see, hear, smell, feel, and otherwise perceive their world. They communicated with one another and found endless ways of making homes and feeding themselves, often developing complex social interactions. They developed marvelous furred and feathered body designs, many with striking colors and seductive dances to attract their mates.

Where could Gaian creation go next? What was possible that had not already been developed?

O   O   O

Think about the world as it was with this tremendous variety of monera, protists, funguses, plants, and animals woven into and weaving the patterns of life. Rock had transformed itself into countless creatures, which had created a rich atmosphere, nourishing seas and soils, producing vast sea and land ecosystems for themselves. Every species of Gaian creation had its part in the dance. Their co-evolution contributed enormously to the overall evolution of their entire planet body.

This great Earth being knew itself in the same sense that we have talked about our bodies knowing themselves. It had and still has the body wisdom to take care of itself and to keep on evolving. In particular, it had learned a great deal through so many generations of creatures living out their lives in concert—in cycles of individuation, conflict and negotiations to new cooperative arrangements.

All this went on without those creatures thinking about their lives or relationships with one another any more than do our own cells and organs ponder the meaning of *their* lives.

Like the cells in our own bodies, they played their parts well. Their perceptions and communications were good enough so that even if they suffered from conflicts and had to learn to resolve them through all sorts of trial and error, the one thing they did *not* suffer was confusion. In a way, they were like actors who played their parts very well even when improvising—actors who said their lines appropriately. But could they have imagined what the whole play was about in the ways *we* try to figure it out?

This seems to be where the next step in the dance came in— with the evolution of a very complex and flexible big brain, in a flexible body with a variety of sensing devices, dextrous hands including opposing thumbs, and a throat with vocal-chords uniquely suited to speech between lungs and mouth. This would be a brain capable of finding external ways to communicate with other such brains through language far more complex in expression than body gestures, grunts, roars and whistles could achieve. Through the social use of language, the new big-brained creatures were to become more overtly aware of themselves and one another and of the play they were part of—to wonder about it and about their roles within it.

The biggest-brained mammals in the sea were the whales and dolphins. And what successful creatures they were! Their bodies had evolved in perfect harmony with the sea to which they had long ago returned. They could roam the watery three-fourths of Earth's surface, seeing by sonar when the water was murky and with their eyes when it was clear. They could adjust to the coldest and warmest, the deepest and shallowest waters, and they could devise a language in which to talk and sing to one another from one side of an ocean to the other.

It is quite possible that cetaceans—whales and dolphins—have evolved the ability to think about their world and to share their thoughts with each other. Dolphin language, for example, is very complex and apparently very much faster than ours. We have just begun seriously to study it, and so far we have made little progress in understanding it. We have even less understanding, or even comprehension, of what their telepathic communications may be like. All we do know is that if brain size and anatomical complexity are clues to intelligence, they outstrip us.

Until very recently we humans believed ourselves to be the only intelligent creatures, and we didn't treat other creatures with much respect. In fact, we have come close to killing off the whales and dolphins just as, earlier in our history, we apparently killed off most species of elephants—the only other land mammals with brains bigger than our own. *Homo sapiens* may even have killed off other species of early humans while competing with them for food. The last other known human species was Neanderthal, who disappeared from the evolutionary record only about forty thousand years ago. Some scientists think they may have disappeared through interbreeding with our own species; others think we killed them off.

In any case, cetaceans had little to worry about until our own human species evolved brains as big and clever as their own, thirty million or more years after theirs had evolved. Though we humans evolved so much later than cetaceans, by Gaian standards we evolved with incredible speed, our brains blossoming suddenly, expanding in size almost explosively from the much smaller brains of our apelike ancestors.

Stories abound about our human origins, from the better known religious stories to no few scenarios in which we result from inter-breeding between extra-terrestrials and apelike Earth creatures, or from the takeover of the apelike creatures' bodies by more evolved non-physical beings. While we have, at least as yet, no clear confirmation of any such stories, we would do well to recognize that the purely Earth-evolutionary story of human evolution still contains plenty of gaps and mysteries. With those limitations in mind, let us continue the story as scientists have pieced it together.

To become human, those ancestral apes would have to do something similar to what their much more ancient polyp ancestors did. They would have to practice neoteny—that is, they would have to remain childlike by not growing up into mature apes at all!

Piecing our own history together from fossils and other biological evidence, we get a story of human evolution goes something like this: Generation after generation, some baby apes were born prematurely, until at birth the bones on the tops of their heads were still soft and not grown together. This permitted their brains to grow much larger *after* being born through the limited diameter of their mothers' pelvic bone door to the world.

In the first year of a human baby's life its brain grows three times bigger than at birth, quickly passing the size of a grown-up chimpanzee or gorilla brain. But human faces keep a babyish flatness instead of developing grown-up snouts and jaws and bony eye ridges. If you look at chimpanzee families, you will easily see that people, even grown-up people, look much more like baby chimps than like adult chimps.

Scientists have shown that our DNA is 99 percent identical with that of chimpanzees, and have estimated that we branched off from common ancestors only two or three million years ago.

Some scientists find it difficult to believe our species can be so different from our large ape relatives with so little difference in genes, but perhaps the major changes in our bodies have been due simply to a few genes that regulate maturation. The potential for greater size and complexity in ape brains may be there, but locked up by the early sealing of their skull bones.

However humans came about, it is clear that genetically we are very closely related to our much more peaceable chimp and gorilla cousins, and this may help us overcome the obsessive idea that we are naturally violent creatures. On the other hand, eager to see our own violence as natural—and at the same time to see ourselves as better than our ancestors—we emphasize the occasional violence we *do* find among great apes, as well as among pre-modern peoples.

Our obsession with violence is worth a brief digression. We teach history as the history of warfare, naming its great ages after metals from which human males made weapons. We teach *his*tory, rather than *her*story, though half of humans are female. Schoolchildren, asked to tell what they know about Incas and Aztecs, are far more likely to remember warfare and human sacrifice than their phenomenal agricultural sciences and architecture, their metallic arts and weavings—as if such great economic and artistic cultures could have been built on mayhem and murder. Of course we do not teach children that European culture at the same time was built primarily on the massive human sacrifice of the Inquisition, which was also Europe's chief export to the Incas and Aztecs, in turn for their gold and silver, which financed Europe's industrial revolution.

All this to say that questions of violence are truly of huge importance for humanity, but social bias in discussing them clouds the real issues. Violence is not 'good if ours and bad if theirs'—it is a danger to all. The really important issue is that we endanger ourselves as a species by believing we cannot evolve from the violence we do to each other and to other species—the competitive belligerence from which we must evolve to peaceful cooperation as so many other Gaian species have done.

Back to our story, the babyish new apes, without big snouts in the way, could easily see what they were doing with their hands, which had long been good at grabbing and holding on to things. After all, they had been living a life of swinging through trees for millions of years. Now, more and more, perhaps for lack of adequate food, these evolving creatures foraged and made their homes on the ground instead of up in the trees. Their arms and hands, no longer so engaged in swinging, were free to do new things. In time, their opposing thumbs grew increasingly dextrous at holding and making things. Meanwhile, their necks and hipbones, legs and feet, were gradually re-patterned for walking and running upright.

The longer these new upright mammals stayed babyish, the later their teeth came out and the less hair they grew, though they may also have lost hair in evolving a cooling system that let sweat out all over their bodies. This cooling system helped them run for such a long time that the animals they learned to chase for food and clothing tired out before they did.

Young humans needed warmth, protection, and affection. They needed to be taken care of much longer than other young mammals. But their long childhood gave them time to play and be curious and learn new things, and parents had plenty of time to teach their children before they grew up. In fact, it is because we never really grow up the way apes do that we can keep on playing and learning new things all our lives—providing we don't seal our own brains up with fixed ideas as inflexible as the old innate behavior patterns.

There were endless new challenges for these first humans. Without the warm coats, strong claws, and long teeth that other animals their size had, they had to survive by using their big brains, their clever hands, and their ability to run longer, if not faster, than many other animals. But perhaps this was in their favor, for even today the most creative and successful people are often those who didn't have everything they wanted early in life and had to work hard to overcome disadvantages.

Early human diets included a wide variety of leaves, roots, seeds, nuts, fruits, eggs, grubs and easily caught small animals, including fish and shellfish. Nets and snares for birds, fish and rodents were probably the primary hunting and gathering tools for a long time. But our ancestors also learned to drive successful predators away from their game catches and acquired a taste for hunting such game themselves. No doubt this made them more inventive. Tools and weapons extended the use of their hands; keeping warm by turning the larger animal skins into warm clothing for themselves was as important as eating their meat.

We can be quite sure that that early humans were both cooperative and scrappy, with emotions from avarice to anger, from lust to laughter, from fun to fear, playing important roles in determining their behavior. Families and groups of families shared caves and other shelters as well as their ways of life, living together where food was plentiful, traveling together when food got scarce. As they became ever more social and communicative, we can imagine them laughing and dancing with joy when life was good, growing frightened and hostile when food was scarce, when dangerous animals prowled near, or when fires were started by lightning near their homes. Yet there is no reason to assume they were any more hostile to one another than chimps or gorillas are. Even if we became a predator species through hunting, it would have been highly unnatural for us, as a mammalian species, to kill our own kind. The question may be, whom did we recognize as our own kind?

It's easy to imagine how exciting it must have been when truly new improvements were invented. Who were the first people to carry burning sticks to their caves, learning to feed flames without letting them spread, learning to cover coals so they could be re-lit the next day or carried to a new home. How proud they must have been to teach such discoveries to their clan or tribe.

Like apes, humans had very flexible, expressive bodies and a natural talent for imitation. Human language probably began with dances and making decorative marks on their ever more naked bodies, to show each other, and later remind each other, of their experiences. Wearing clothing gave opportunity for elaboration into special costumes with headdresses and adornments as they created these rituals. We can imagine them dancing their hunting adventures, copying animals in courtship dances, and so on.

We can also guess that the sounds they made as they danced took on meaning, eventually becoming spoken words that were symbols for actions and things, just as their drawings on rock or on the ground became picture symbols. Once there was spoken language, it must have been much easier for them to learn and teach ever more complicated ways of life to one another.

Quite as the ancient bacteria had gotten together to do different jobs within the same cell walls, and as the cells of protist colonies, or the ants of ant colonies, divided their jobs among themselves, so humans now organized themselves into communities where different people did different jobs. Some hunted or fished; others scraped skins for clothing. Some became specialists in making tools or baskets or clay pots, which were hardened in fires. Some were no doubt better than others at drawing, dancing, or telling stories. And some became leaders—wiser elders, chiefs or medicine people—who organized jobs, drew up rules to live by, and made decisions on what to do when others could not agree.

Human communities, as they got larger and more complex, evolved leadership and governments just as eukaryote cells had evolved nuclei and just as animal bodies had evolved brains. Every holon that grows larger and more complicated must evolve some way of organizing itself to simplify and manage its complexity if it is to survive. When human communities got too large, they split up, or budded off, into new colonies, thus reproducing themselves much as ancient bacteria had.

O   O   O

Among the oldest artifacts of Stone Age societies are motherly female images carved in stone or modeled in clay. Some of the oldest stone-age temples we have found were actually built in the shape of human female bodies, such as those in Malta, or in mounds that symbolized them, with internal womb chambers and birth passages, as in Newgrange, Ireland. So many such images and temples have been found by now, without any male images from the same times, say of hunters or warriors, that it is clear their makers were more interested in the giving of life than in the taking of it.

The giving of life is actually a lifelong business of feeding and otherwise nurturing people. Early humans probably began agriculture simply by leaving pits and seeds along their habitual tracks, as Amazon hunting cultures still do today. This rearranges ecosystems favorably for humans—a kind of intermediate stage between gathering and gardening. Gradually they learned to grow reliable food supplies in their settlements and to keep and breed captive animals. Where climate and soil permitted, they organized themselves into villages with fields. The art of agriculture evolved from simply planting seeds to selecting the best seeds from each harvest for planting and giving as gifts, as well as to preparation of food for storage against times of need. As humans cleared larger and larger stretches of land, they began to seriously alter their environments, eliminating some species to nurture others, changing the genetic identity of plants and animals through their selection.

The agricultural revolution of the Stone Age was no doubt the greatest transition in human history, forging our destiny as a species that would change the face of the whole planet, destroying and rearranging things to our desire. Evidence suggests it all began peacefully and modestly with simple agricultural techniques for growing plant foods and keeping herds of animals for milk and meat and wool. In nurturing one another, people learned to spin yarn and weave cloth, to mold clay into vessels, to make houses of earthen bricks and furniture of wood. They learned to smelt metals from the earth for making images of worship, adornments to wear, and stronger tools, as well as for weapons.

They made musical instruments and danced and sang in worship of deities as well as to celebrate and tell colorful stories about their world and themselves. Some of those stories were reports, others were used as lessons, still more were told just to exaggerate, boast and entertain. We assume it was difficult for early people to explain their world to themselves—that nature was an incomprehensible and unpredictable power to them and death a frightening mystery—but this is not really likely. Today's surviving indigenous cultures—tribal cultures still living on their ancestral lands—have a much closer relationship with nature than does our technological society. Those that have been able to maintain their ancient ways despite all efforts by conquerors to destroy them, tend to see themselves as active and responsible co-creators of their ecosystems, something our dominant culture needs badly to learn from them. More on this in Chapter 19.

When we continue the story of human social evolution in the Stone Age and after, we will see that sustained violence among humans was quite likely a distortion of our humanity that came fairly late on the scene of human civilization. This should give us hope that our violent phase may be a temporary aberration from which we may recover. But let us first take a look back and reflect on the fact that for the greater part of our existence we were not so different from the rest of the animal world.

O   O   O

Looking at ourselves today, it's hard to see humans as natural animals. But from Gaia's perspective that is just what we are—and a very new animal species at that. For most of our history we did little more than mimic adults as children, mate and give birth, shelter and care for our young, gather food, defend ourselves and our homes against other animals, then rest or play with one another before we fell asleep to dream our restless dreams. Compared with the few million years in which human life was not very different from that of other mammals, our civilizations are still a very new development in a still juvenile species.

Imagine squeezing the four and a half billion years of Earth's existence into just twenty-four hours. While bacteria were born long before dawn and had the world to themselves until midday, humans came on the scene only for the very last minute before midnight! And only in the last *second* of that minute have we formed settled societies we call civilizations. We really have just begun to become fully human. Consider the experience whales have gained being already in their present form for some thirty millions of years.

Since people evolved—however recently by these timescales—we have survived many ice ages, which were believed until now to have been abnormal times for our planet. Lovelock and others presently working out Gaian physiology suggest that ice ages may be the Gaian norm during phases when continents are separate and spread out. Between these phases the Earth seems to warm up, creating periods between the waves of ice ages that are abnormally warm, like fevers. We now know of seventeen distinct ice age phases, most of which occurred before humans evolved.

Ice ages are only a few degrees of temperature colder on the average than the times between them, though we sensitive creatures consider them great extremes. In any case, during ice ages, great patches of ice move, as we know, from the poles over parts of the Earth that are warm between ice ages. As more ice is produced, the sea level is lowered and more land is exposed near the equator—as much land as there is in the whole continent of Africa. When the ice recedes again, these areas are flooded, but ice-covered areas warm up, melt the ice, and produce rich new forests and grasslands.

Humans were apparently driven ahead of the great sheets of ice when they advanced, and humans later followed in their wake as the ice sheets receded. They wandered to places that were rich in food and water and settled down until, many generations later, a new wave of ice drove them back. By the time the last ice age was over, around ten thousand years ago, people had spread themselves out over much of the Earth's land, taking with them their early civilizations, their ways of changing the land, their seeds, and their stock. But before we go on with their history, let's look a bit more closely at the spectacular brains that made all this possible.

11

The Big Brain Experiment

If you could look inside your own head, as we looked in on our developing embryos earlier, you would see yet another kind of evolutionary record. The innermost part of a human brain looks much like a reptile's brain, and it seems to be this deep core of the brain that sometimes makes us huff and puff and attack others automatically, as though we have disconnected it from the rest of the brain, which might give thought to what we are doing. Wrapped around this core, is a more recently evolved part of the brain, which, together with the core, looks like a more modern mammal brain, such as that of the horse. This mammalian complication of the brain permits those mammalian feelings we call anger and love, sadness and joy. It also appears to make us playful, curious, and eager to learn.

On the surface of the human brain we find the newest complication—the *neocortex*, or 'new bark' growth, which ripples and folds itself to look like a great elaborate walnut inside its skull shell. Despite some success in mapping it, the brain is not separable into parts such that we can say that this or that feeling or behavior has its locus in a particular place. Yet we know it is the evolution of the neocortex—richly interwoven with the inner brain—that permits the whole brain to demonstrate our special human abilities. We can remember our past in detail, compare it with our present, and, on that basis, make plans for our future. The neocortex is very much involved when we create our ideas of the world, communicate in language, think up our inventions from courts of law to computers, create works of art, research scientific questions, decide what is good and what is bad, learn and think about our relationship to all other creatures of the Earth and even to the whole universe.

There is a great deal that we still don't understand about brains. We can study their evolution and construction, count their cells, record and measure their patterns of chemical and electrical activity, and yet we do not really know *how* they do what they do. There is a very good chance that our explanations of them will change dramatically in the future. Just for example, if physicists are right in saying that fundamental reality is not material—that matter arises through a continual transformation of cosmic consciousness—then brains may be material devices permitting consciousness to operate in, or to interface with, the rest of material reality. This view would see brains as biological devices created and operated by consciousness, rather than seeing them as biological devices which give rise to consciousness, as has been supposed—a rather complete turnaround!

It is wise to remember how much our western scientific stories have changed in the past century and to realize that they are likely to change even more in the next. The concept of cosmic consciousness is still new in western science, though earlier eastern sciences saw it as fundamental to the universe and were very interested in distinguishing it from the kind of consciousness we are familiar with from our ordinary waking experience. Cosmic consciousness can only be directly experienced through long training in meditative exercises, and is most usually experienced as a blissful unity. Ordinary waking consciousness, on the other hand, is always present, usually complex, and seems to have evolved through the evolutionary experiences of Earthlife. What we call the subconscious seems to exist somewhere between the two, and reveals itself in the dreams we spoke of earlier.

The philosopher Alan Watts suggested the universe might be a great game of hide and seek played by God, who was everything and so had to play the game alone, hiding in rocks and trees and people, waiting for them to discover who they were. We might also see the universe as a vast learning experience, with ourselves currently at its leading edge, trying to figure out how it all works by looking back on it through our unique kind of consciousness. As we evolve and learn, so do our stories change, whether they are religious, scientific or other.

It is as though cosmic consciousness keeps trying out new possibilities through biological and social evolution. When we compare the brains of many species, we can see quite clearly the parallel evolution of ever more complex brains with ever more complex behaviors. Species consciousness, communication, and freedom of choice in behavior seem to have evolved gradually, becoming most complex in the relatively bigger-brained species. In our own species, and perhaps earlier in cetaceans, there has been an unusual explosion of these talents. Like humans, dolphins seem to have ideas, ethics, complex languages, and personal names. While scientists have not succeeded in finding a language bridge between humans and dolphins, there is an increasingly large literature on telepathic communications between people and cetaceans, including repeatable experiments and evidence of dolphins healing people.

Indigenous people share an understanding of nature as a vast and continuing dialogue among all parts and species of Earth—as one great family—with humans as the youngest member with the most to learn. Many of them consciously participate in this natural dialogue as co-creators (see Chapter 19) but a branch of humanity that chose to cut itself off from this dialogue now dominates Earth. Enamored of our big flexible brains, we of western culture have—in our minds, if not in physical reality— disconnected ourselves from the community of life. We study it as though it were separate and work to control it for our own purposes.

No other species has been in a position to do this, for none is so free to choose its behavior and thoughts at every turn. None is so clever in inventing and producing technology, so powerful in its ability to kill or protect other species, to destroy or preserve whole ecosystems worldwide, *or* to pretend it is separate from the rest. Truly we are a Gaian or universal experiment in freedom, and a risky one at that.

O   O   O

Most animals, as we saw earlier, do most of what they do as innate behaviors—that is, without having to learn by trial and error. There is little you can teach an ant or a lizard, as virtually all the ways such a creature responds to its environment seem to have been built into it by its evolution in community. Their behavior is carried out by their nervous system and the rest of their physiology in response to their changing environments— the things and events they encounter. In the last chapter we saw that the larger, more complex brains of mammals freed them from some of their rigid innate behaviors and let them learn new behaviors through feelings and experience. In mammals, innate behavior is still apparent, but feelings and learning and voice communication add variety and richness to behavior.

There are a few things we humans do automatically, as we say 'by instinct,' such as running from danger or attacking when we feel threatened, seeking mates, feeling love for our babies, and seeking food or a place to lie down when we are hungry or tired. We don't have to think about such things. But we have no innate programs for painting pictures or digging up fossils or building airplanes or philosophizing or designing hospitals or doing the millions of other things we do.

Our behavior is guided partly by basic needs, a great deal by what our society has taught us, and somewhat by personal choice. What we do by our own choice usually depends on feelings and experience with our past choices, some of which become habit. What we learn from our society depends on how other people's feelings and experience with past choices have been turned into customs or social habits and rules. It is our society, for instance, that sends us to school and tries to keep us out of trouble with others.

Let's consider for a moment this matter of staying out of trouble with other members of our species. In most other social species, built-in behavior patterns keep individuals out of serious trouble with others of their kind. Many species of reptiles, birds, fish, and mammals, for example, make their homes in places or territories they have claimed as their own. When other members of their species come into this home space, the resident inhabitants warn them with species-specific dances or songs. Usually the intruders back off and leave. Even if it does come to a fight, the intruders know innately to give up before being seriously hurt or killed. You may have noticed that the loser in a dog or cat fight turns up his or her throat to the winner, a behavior kittens and puppies show even in early play. At this "I give up" signal, the winner's attack stops as if suddenly switched off.

Animals use ritual dances and fights to win and protect their homes and mates as well as to raise their young in safety, providing they have adequate territory. The rituals work like a system of rules for living together in reasonable peace—rules of behavior drawn up in evolution as much as was the structure of their bodies. These rules help them balance their lives by spreading each species out over an area of land or sea without crowding, so that all have an adequate chance of getting enough food. In this way, what we call innate territoriality and aggression work for the good of the individual as well as for the good of the whole species, with relatively little aggression compared with our human use of it.

Just as animals know innately how to share land without killing one another over it, they know innately when they have enough land and food for their needs. An animal may hoard just enough food to get itself through a hard winter, *but no animal except the human one piles up food or takes land beyond its need.*

The price of our freedom to decide our own behavior is the loss of such innate rules to limit our own aggression and greed. Like other animals, we have an innate urge to supply ourselves with the necessities of life, to win mates, to make homes for our families, and to protect ourselves and our families against intruders. But, unlike other animals, which know innately just how to act on these urges, we are free to act on them in countless different ways. We must make our own rules for sharing or not sharing the Earth's resources with one another and with other species. If we share resources fairly by common agreement, there should be no reason to use aggression. Aggression is not something piling up in us that demands an outlet, as some psychologists and sociobiologists have suggested—rather, it is a reserve capacity available in situations of real need. Whether such need arises is almost entirely up to us.

As we big-brained mammals lost the rigidity of innate behavior and gained freedom of choice, we also gained our unique kind of consciousness—our reflective awareness of what we are doing, our memory of what we have done, and our projected images of what we might do with our awareness of choice. This conscious awareness that we live in a linear past, present, and future, in which there is cause and effect, makes it possible for us to predict on the basis of past experience what the effects of our behavior will be. Even though physicists now tell us—as eastern philosophies did earlier—that cause-and-effect spacetime is an illusion, this kind of perception of our reality serves as our *guide* to behavioral choice.

Many humans find that with spiritual practices such as meditation and contemplative prayer, they can directly perceive cosmic love and unity through inner senses. This adds a very important extra dimension to their behavioral guidance system, because they perceive all things—including people—as One. The great teachers of our world, such as Jesus, Mohammed and Buddha, gave us ethical systems based on this kind of perception—ethical systems showing such universal interconnectedness that what is done to others is done to yourself. Now western physicists showing us non-timespace and non-locality begin to teach the same thing!

The problem in our dominant human culture is that we are using only a linear perspective to predict only the very short-term consequences of our behavior, while failing to consider the broader and the long-range results. Most of us know by now how some so-called '*un*civilized' native cultures taught their people to think of the consequences of their choices for seven generations ahead, but we have not yet adopted the practice. In other words, we have not yet learned to use our conscious freedom of choice for the good of our whole species over time. Rather, our so-called 'civilized' history—at least the most recent five thousand or so years of it—shows that humans have used opportunity and power again and again to take much more than they need, usually by taking it away from other humans and killing them if they resist.

Western society, priding itself on its enlightenment, has continued this grim record, prompting Gandhi to respond to the question: "What do you think of western civilization?" by saying he thought it would be a good idea! Clearly, we lack the built-in limits of other species and must choose to limit our aggressiveness for the health of our species, if not for our spiritual development.

There is now good evidence that many early agricultural societies were indeed peaceful, sharing land and other resources, not making war on other societies. Eventually they seem to have been taken over by certain unsettled nomadic peoples who had become aggressive, perhaps for lack of adequate resources. Since that time, there have always been some people who have far more than they need and others who have far less. We simply do not share resources as well as most other species do. But, then, we are still very new, and as we will soon see, there are signs that we may be working this problem out.

Our need to live in societies is as much our natures as our territoriality and aggression. But here, too, we are very free to decide just how to act on it—what kinds of societies to create for ourselves. Insects such as bees, ants, and termites evolved highly organized societies, as had earlier bacteria and protists. We do not know whether they evolved as societies, similar to cell colonies, or whether they evolved first as individuals and later assembled into societies. In any case, social insects build whole cities, make farms to raise plants and other insects for food, have queens, soldiers and workers who tell one another what to do by chemical messages, make wars, and capture slaves.

We are astonished to see the social insects doing so many things that seem so human to us. But actually they could hardly be more unlike humans. Social insects have been living these complicated social lives for millions of years in the same old way, for their hard, external skeletons kept their brains very tiny and unable to evolve. The things they do all their lives, generation after generation, are innately determined. A worker ant cannot change its mind and become a soldier; a queen bee cannot change her mind and run things differently.

To compensate for the smallness of their brains, these insects specialized their functions in such a way that the different special-ists together form one social body. Social insects need one another and cannot survive as individuals alone. It's no use trying to keep a single ant as a pet, for it will lose its appetite, get sick, and die. In a way, it is not a whole creature in itself, but an organ in, or a part of, its social body—its anthill society. In a way, an ant is to its society as a mitochondrian is to its cell.

Mammal species, such as our peaceful gorilla cousins, have much less rigid social behaviors than ants, and yet they, too, have innate behavior patterns that preserve the societies they are born into—social structures that have been tested in evolution and have proved healthful for the species. It is interesting to observe that very aggressive species, such as certain baboons of North Africa, have a very rigid social dominance system, while peaceful species, such as the Bonobo chimpanzees, show greater equality and opportunity to change roles.

As we will see later, we humans are free to form and test and change our own social structures, and indeed we have tried many different kinds of societies in the course of our history. Yet, for all our experiments, hardly any humans except a few vanishing indigenous peoples today live in a social structure that is truly healthful for all the people living in it as well as for the other species among and around them. But again, we are still very young, and there is every indication that we could solve this problem by understanding our living planet, our indigenous survivors, and by looking to our remote but peaceful past and truly desiring a peaceful future.

O   O   O

Our big free brains and our clever hands have permitted us to make dramatic and sudden changes in ourselves and in our world during the past few thousand years. While this is merely the blink of an eye on the time scale of Gaian evolution, it is long enough for our chosen ways of life and our inventions to have become at once a threat and a promise to our whole planet. We may now have become as dangerous an experiment in life forms as were the blue-green bacteria—recall that they learned to make energy from sunlight and eventually covered the world with poisonous oxygen, killing off countless other bacteria. Still, we humans *could* prove to be as much a step to healthy new Gaian developments as were the breather bacteria that found a way to use the poisonous oxygen in the most efficient energy-making process ever invented—the breathing process that permitted us to evolve.

We are not actors learning our parts in our sleep and playing out our lives unaware. We are awake and free to learn what the play in which we are players is all about, and we are free to change the play by our own choice. We live right now at the most exciting time in our history. It is a time in which we can see ourselves as never before and understand who we are through knowledge of our own history, our evolution, our universe. Perhaps most importantly, we can see what children we still are as a species and what opportunities there are for us to grow up.

If we evolved by refusing to grow up as apes, then sooner or later we will have to grow up as humans. And to grow up as humans we will have to take the responsibility for using our freedom in healthful ways, to help rebalance the great ongoing dance of Gaian creation and to develop harmonious new patterns within it. But to understand just how we might go about this, we need to look at human history not as something separate and different from biological evolution, but as a continuing part of it—as the social evolution of one species within the Gaian life system.

What, then, has been the historical evolution of roving human hunters and settled human planters in the period of known civilization, which now extends back some thirty thousand to forty thousand years?

Some surviving groups of indigenous peoples living their traditional lifestyles demonstrate the kind of ecological harmony we see among other species in mature ecosystems. But, as we said, the rest of us diverged from this path thousands of years ago. People in settled agricultural societies, despite generally favorable climates and ways of providing for themselves, surely faced hardships such as floods and droughts or epidemics of disease. Such challenges made humans ever more inventive in their lifestyles, just as similar challenges had made ancient bacteria inventive in theirs.

People learned to store food against times of need, to make rules for sharing land and working it, to make medicines from plants, to make canals to bring water from rivers to fields, and to build boats to carry things up and down rivers and coasts for trading with other peoples. Archaeologists, in studying early human civilizations, are now finding more and more evidence that early agricultural societies everywhere worshipped a Mother Goddess as the giver of life and regarded men and women as equal partners, though the actual management of these ancient economies may have been, on the whole, the responsibility of women. Such civilizations developed agricultural techniques, other arts, law, and trade over as many as forty thousand years of peaceful evolution— by far the longest part of known human history. We shall return to them in chapters to come.

Four or five thousand years after the last ice age ended and about the same number of years before Christ, the goddess-worshipping cultures, especially in the Middle East and around the Mediterranean, began to be conquered one by one. Armed tribes of unsettled nomads and hunters on horseback came in waves from less plentiful colder or desert climates, searching for a better living. Ruled by men and worshipping Father Gods, these tribes broke into the more peaceful agricultural settlements, conquered them and, in many instances, stayed to form new more complex social orders, which eventually grew into warlike kingdoms or empires. Until recently these kingdoms were considered the cradles of civilization, as we did not know of the earlier peaceful civilizations.

Again we are reminded of the primeval Gaian world, where hungry monera forced their way inside other monera to get at their rich supplies, then stayed and multiplied, eventually shifting from competition to cooperation as they formed the much larger and more complicated protists. Perhaps without the invaders, the settled human cultures would have remained like bacteria cloning themselves—producing the same peaceful offspring cultures over and over. Instead, the conquering tribes came in like the invading monera eons before, taking over their hosts to pursue their competitive interests, staying to build their own empires. If this pattern among humans follows the pattern of the ancient bacteria—and we will see more signs that it does—then we, too, will work out peaceful cooperation to replace our competition with a healthier life for all.

In Gaian evolution the cloning monera took billions of years to change themselves and their environment in ways that permitted the much faster evolution of oxygen-breathing protists and larger sexually reproducing creatures. In human social evolution the goddess-worshipping cultures took tens of thousands of years to develop agricultural ways of changing their environment to support themselves, while the later god-worshipping cultures have changed themselves and their environments tremendously in only five thousand to six thousand years.

Clearly human social evolution sped up and created more varied and complicated patterns from the time of these invasions. Unfortunately, however, the competitive, exploitative situation that went on inside large bacteria before they became nucleated protists is still going on in the human world. The male-ruled conquering tribes took almost all women's social power and status from them, declaring them inferior and setting up other inequalities in society. The records of these invasions are the first clear indication of large-scale violence among humans.

After the invaders conquered these very differently organized societies, they imposed their own social structures and customs upon them, though some of the old ways no doubt persisted in the new hybrid cultures. We humans are creatures of strong habit; our cultural rules, beliefs and rituals are the glue of our societies. Our very ability to function has always been heavily dependent on our social ideas and structures, and we have therefore fought to preserve them.

As the new larger cultures fought wars with one another, the losers were absorbed into the winners' empires and forced to abide by their customs, though some always persisted and modified the dominant culture. Thus empires grew ever larger. Great empires were formed by Sumerians, Assyrians, Etruscans, Babylonians, Chinese, Egyptians, Persians, Greeks, Romans, Aztecs, Incas, and Mayans in South America; Kush, Nok, and Axum in Africa.

Within each empire, people were kept organized by rulers who laid down laws and kept guards and armies to enforce order and fight wars. Wars brought captured slaves, or brought whole cultures into the empire, though some cultures were absorbed peacefully. Recall how bacteria and larger creatures ate other bacteria or creatures that became parts of themselves.

Often such rulers maintained their power by claiming to have been chosen by the gods; some even said they were gods themselves. Some were benign, others less so. Most of the ordinary people in these societies were workers who grew crops and husbanded animals, hauled or channeled water, mined metals and precious stones, made pottery, tools, and weapons, built cities with huge, beautifully decorated palaces and temples for their rulers, and otherwise transformed the natural environment to human use.

Empire building by male-ruled class-structured societies became the main process and pattern of human social organization, and in one form or another it has continued right up to our present time.

In ancient Greece a brief experiment in limited democracy was made as collective rule by all non-slave male citizens. Truly inclusive democracy—from *demos*, which means people or community, and *kratos*, meaning government—is something we are still working to achieve. Later we will look more closely at this experimental male democracy and why it did not last, though it powerfully influenced our whole modern world. For now let us move quickly through history to see its main pattern, to see how we humans used our big brains to continue the task of empire building.

The Roman Empire conquered Greece and later evolved into the Holy Roman Empire, which ruled Europe. The Byzantine, Chinese, and Ottoman empires were formed in the East; each of them falling in its turn, and eventually the human world divided itself into countries, or nations, as we know them today. But very soon the most powerful of these countries began building their own empires by conquering new territories far from home.

Two human inventions—the compass and the printing press—helped to expand empires across oceans all over the planet and to spread the knowledge and culture of empires over almost all of humanity. Soon more machines were invented to make things other than books in large numbers—the age of mechanical industry had begun. The word *manufactured*, which literally meant made by hand, soon came to mean machine-made.

O   O   O

The making of things by machine transformed the whole human way of life, bringing about a new kind of empire and building a new road to riches. The biggest empires at the beginning of the industrial era belonged to the kings and queens of seafaring European countries such as England, Holland and Spain. With ships and weapons they conquered peoples in Africa, Asia, Australia, and the Americas, carrying off riches and rich natural resources, making the people slaves wherever possible, and taking over their lands as colonies. The Europeans used these riches to develop their new industrial way of life. Native peoples in Europe's colonies were forced to mine iron, copper, and other metal ores needed by the conquerors to build machines, weapons and transportation systems. They were also forced to give over their agricultural diversity to grow large single species *monocultures*—food crops, as well as rubber, tobacco, cotton, and wool. These were then exported to Europe, where other poor, landless people worked at the machines that turned the crops and raw materials into products for sale and who mined the coal that fueled the machines.

In time the colonies began fighting for and winning their independence, thus breaking up the empires. The North American colonies were first to win their independence, and they quickly developed their own machine technology and industry to compete with European empires. The success of the United States, after it won its independence, however, was not typical. In most other colonies, the best land and resources had been taken over by European settlers who had gotten rich by shipping raw materials to the mother countries. After independence was won, these countries continued the colonial way of life, exporting raw materials and food crops to industrial countries and importing machine-made products.

Many peoples who had once cared for themselves independently by hunting for or growing everything they needed to live on their own land are worse off in their now independent countries than they were before the colonial empires were formed. While they are regarded as backward by Western standards, they have actually been systematically underdeveloped—prevented from proceeding with their own natural development in order to support already wealthy countries. Even after they win their independence, the people are forced to continue working the land for others, as they did in colonial times, often farming a monoculture crop or mining a single metal ore or fuel for export. They must buy their food, clothes, and housing with what little money they are paid, so they cannot escape poverty and often suffer hunger and illness.

Such peoples have lost not only their land and natural food supplies but their whole way of life as well—their tribal organization, their nature religions, their arts, and often even their languages. Progress, they were told, meant learning the ways of Europeans and Americans. But though they gave up their old cultures to adopt these new ones, progress for them only meant getting poorer in every way while those who owned their land got ever richer. Later, the richer countries, through international organizations such as the World Bank and the International Monetary Fund, lent these struggling countries money for development. Most of this development served the lending countries more than the developing countries themselves, yet it put them ever deeper into debt, creating a huge problem of global inequity we call 'haves and have-nots.' Imagine such a problem among the cells of your body.

Industry changed the way of life in powerful countries as much as in poor ones. The rich were no longer those who owned big portions of undeveloped land, but those who owned machines and bought up land, stripped it, and put it to use in producing raw materials for industry. More and more people were forced to give up subsistence farming and go to work in factories, swelling cities around the factories as the urban way of life became the social standard and the symbol of progress.

Before the Industrial Revolution all humanity lived off agriculture, and few people were rich enough not to have to produce their own food, clothing, and shelter. In today's world, most of us buy almost everything we consume, which has been made, often far away, by others. Even when products are made in our own country, the raw materials in them very likely came from another. The whole world is now tied together by its *economy*—a word coming from *oikos*, meaning household, and *nomos*, meaning law or management. The human household, once a local family or tribal unit, now encompasses the whole world. A vast web of transportation and communications lines has been spun around our planet to move about raw materials and finished products and people to manage the global household.

In just a few hundred years, then—much less than the blink of an eye by Gaian standards—our brash young human species with its big brain and clever hands razed vast natural ecosystems to transform them into a single economic empire covering the whole planet and ruled by the rich industrial countries. Yet during two World Wars and the long Cold War following them, both the rulers and the ruled of this empire were politically split into blocs of countries with conflicting, competing schemes for managing this worldwide *globalizing* economy and its people.

Now, all of this is changing dramatically again. The Industrial Era has given way to the Information Age in the evolution of globalization, and its most important invention—the Internet—is forcing us to understand ourselves as a single living system, a body of humanity. We will see this in greater detail in Chapter 16.

O   O   O

Our modern world, with all its successes and problems, seems easier to understand if we look back again and again to the ancient bacterial monera in the process of evolving protists. The exploitation of host bacteria by hungry invading bacteria that needed their resources reminds us of the God cultures invading the Goddess cultures. It also reminds us of our more recent imperialism, in which some countries invaded others and made them colonies. But using up the host colony bacteria's resources killed the host, and so this process could not meet the long-term interest of the exploiters.

This is the lesson we are learning after thousands of years of exploiting one another in our struggle to become a mature species—thousands of years, which are as nothing for such a big evolutionary change. The healthy cooperative system that began evolving when invading bacteria produced energy for the host in return for raw materials is paralleled today as developed countries help their former colonies develop industrial energy. The problem is that we have not yet worked out fair exchanges. The powerful countries still demand more political and material concessions from the so-called developing nations than they would in a truly cooperative system. But then, the ancient bacteria did not evolve into protists so quickly, either. Only when many bacteria of various kinds had become involved with one another inside the same walls did a cooperative system, including a common nucleus, begin to evolve for the benefit of all.

Human countries have only recently found themselves inseparably linked inside the boundaries of our planet, and we are just beginning to understand what this means. If exploitation and hostile rivalry continue at the expense of cooperation among countries, the new body of humanity may not survive much longer. Evolution takes time, but when a natural system has pushed itself or been pushed to certain limits, it can reorganize itself with incredible speed, as the great extinctions teach us. Humanity has now reached such a critical limit. It has also invented everything it needs in order to accomplish a dramatic reorganization into a healthy cooperative body.

Particularly interesting is the fact that bacteria invented communications systems prior to organizing themselves into nucleated cells, and that nucleated cells invented intercellular communications systems before organizing themselves into multicelled creatures. This is how the Internet will play out its enormous role.

Communications systems, which we humans now have worldwide, are prerequisites to the organization of larger living systems. Transport systems for moving about supplies also play a critical role in the actual organization and function of such larger systems, and here, too, we are well prepared in our transport capability. If the big brain experiment is to be a success—if humanity is to survive as a healthy body, as a holon in the Gaian holarchy—we must use our communications, our information exchange, and our transport as parts of a cooperatively organized body. The sooner we recognize that this is our only viable direction, the sooner we will get on with the task.

12

What the Play Is All About

Earlier we said that humans are the only species in the play of life
that can think about and try to understand what the play is all
about. Yet we are just now forming a scientific worldview in which
we understand our world holistically—as a whole made of interconnected parts. We are just beginning to understand how we are
related to all the other players in our planetary holarchy of holons; to
understand that we are new players in Gaian creation—new players
with the responsibility of exercising the freedom of choice our big
brain gives us in ways that will keep the play going for all of us.

This idea of humans as part of one huge cosmic play, with freedom we must learn to use responsibly, is actually not new, but
ancient. The Hindu Vedists and Chinese Taoists understood things
this way, as did Homer and the first Greek poet to write plays surviving to the present—Aeschylus, who lived in the fifth and sixth
centuries B.C. The plays of Aeschylus are all about the role of
humans in their social and natural world—all about the human

task of making responsible choices within the situations and limits set by the world of human society within the larger natural cosmos.

In fact, this playwright's layered cosmos can easily be seen as a natural order of holons in a holarchy. Aeschylus understood how each human choice in behavior affects not only the doer but also the doer's family, society, and the larger cosmos beyond humans. He saw that the extent of our free choice within the natural cosmos to which we belong is the most remarkable thing about us, and his plays are about the questions people must weigh in making their choices, the effort to understand the consequences their choices will have at all cosmic levels.

The ability to think about choice—to make images of our relationship to our world and imagine the consequences of the alternative choices we can make at each step through our world—is the biggest role of our unique kind of conscious mind. The ways in which we picture our world and our relationship to it—our stories of how things are— are our *worldviews*, and these have a great deal to do with the kinds of choices we make in the play of our lives.

To understand what human worldviews are and where they come from, let's begin by considering how other species view their world. Whether it has eyes or not, every creature has some way of *seeing* its world—some way of getting information about itself and its surround. Without such information and the ability to act on it, no creature could function in its environment.

We have said that every creature is a holon within the larger holons on which it depends. To live, it must *know*, in some sense of that word, what supplies to take in from its environment and what to return to its environment. It must do what it can to protect itself from harm and to do whatever else will help it go on living. Even a microbe can tell whether it is in a plentiful environment or not, can tell what is harmful in its environment from what is not, can tell what is useful in its environment from what is not, and so on. Further, it must coordinate all this information to help itself survive. We can call that pattern of information it perceives its worldview—its map of reality.

The point is that some kind of environmental map, or worldview, is as necessary to the survival of any living creature as is its internal knowledge of how to run itself. In fact, as different creatures evolved, different worldviews evolved. The worldview of a microbe is clearly not the same as that of a marsh grass or a mongoose. Every living being has a worldview tailored to its own needs and experience. This is because each creature is a system capable of interacting with its environment through its unique ability to take in information and act on it.

No creature, even with a brain as sophisticated as ours, *sees* what is *really* out there in its world. Our eyes do not photograph a world independent of us. There is nothing remotely like a photographic mechanism in our eye-brain system, nor is there a world apart from that which we create moment by moment within our own consciousness.

Stop to consider this deeply for a moment. Have you ever had any experience outside your own consciousness? It simply is not possible—not for anyone, not even for a scientist. Now, have you ever had any *direct* experience outside of the present moment? You are not alone or strange, for neither has any scientist. This is very profound. *All experience of the world is through consciousness in the present moment.* Everything else is stories and images created by ourselves—including the image of linear time. Consider that you have a mental story of reading this book over a period of hours or days, but are always, always *in* the present moment. This is exactly what the Eastern philosophies meant—that the world is illusion. Now science begins telling us the same thing—that we create reality with our brains from moment to moment, as cosmic consciousness, which includes our individual consciousness, creates our brains from moment to moment. It shakes one up to understand that deeply, for all humans across apparent linear time and cultures have made up stories of their reality. If there *had* been a reality really 'out there' apart from our perceptions and stories, you would think cultural 'descriptions' of reality would have been much more uniform. Instead, we find that every culture believes its own story and no one else's—more on this shortly. It even means that the story of evolution is *just that*—our story of how things came to be, not some ultimate truth.

Why then are we talking about how evolution happens? Because it is not possible to be human, to exist in our perceived reality, without a coherent story of how things are. When we recognize that the *story* is all we really have, then we see that it is truly *essential*. What matters is to create a story that brings us meaning and fulfillment in the world we see as real. The most exciting trend in our world now is that the stories of scientists and philosophers and religious leaders are weaving themselves into one coherent story told from different viewpoints. If scientists understand an intelligent cosmic consciousness as the source of all creation, and creationists call that source God, our stories are not very different.

So, let us continue *this* story to see if it could bring us meaning, fulfillment and peace in the world we see as real.

Scientists tell us that inside our eyeballs, light does strike our retina in a way that reminds us of a photographic plate or film, and that the light pattern does produce a related pattern of nerve signals that travel to the brain. These nerve signals, however, are soon joined by a far greater number of other signals coming from inside the brain itself, combining the brain's own information with the incoming information to produce our visual images. Not like a camera at all—rather, what we then *see* is this complex production of our brains.

The same thing happens when we look at a photograph. The reason the photograph resembles what we saw with our eyes when we took the picture is that our eye-brain system responds in the same way to the scene as it does to the photo that is a mechanical copy of its pattern of light.

Let's *look* at how scientists tell us a frog sees its world. A frog lives mainly on insects it catches as they fly or swim or crawl by. Its eyes and brain and body are an automatic system that has evolved to see and catch bugs. Whenever a tiny dark speck moves across the piece of world the frog's eyes are aimed at, it shoots out its long sticky tongue and pulls in the tiny dark thing. This system works very well in the frog's natural environment, because tiny dark things moving about that way are almost always insects.

We can fool a frog into trying to eat tiny shadows that we move past it, or into actually eating buckshot pellets that we roll past it. The frog does not learn that the buckshot is inedible, but will keep eating it until he is too heavy to move. Its catch-moving-dark-specks system is built into its brain and cannot be changed by experience. Its worldview is not subject to change in a way that permits the frog to learn which dark specks are edible and which are not.

A human baby also puts all sorts of things into its mouth, but for the baby this is a way of finding out what is edible and what is not, how things taste and feel. Its tongue, eyes, ears, and other sense organs form a system that determines and limits the kinds of information it can receive, but the baby is not programmed to put the information together in very fixed ways. The baby must test its world constantly, learning about it through direct experience of it, as well as through what it is taught and told. So, in time, it builds a worldview. From infancy on, our brain tests each new experience against those we had before in order to keep the worldview we are building consistent.

Let's look at another example of species differences in worldviews. Suppose a child is playing with a cat when it sees a bee land on a pretty flower. And suppose that all three—the child, the cat, and the bee—can talk to one another.

"What a pretty pink flower you have chosen," the child might say to the bee.

"What pretty pink flower?" the bee might well ask. "Can't you see I have chosen this flower for its deep red stripes? This kind has my favorite pollen."

"Red stripes?" the child says. "I should think a bee could see better than that! This flower is pink as pink can be."

"I beg your pardon," says the bee, "but it is you who do not seem to see very well! Look here, cat, is this flower striped or am I crazy?"

"Both of you are nuts," says the cat, "making a fuss about a flower. They all look alike to me, and rather dull-looking things they are, sitting about as they do. Now a grasshopper is another matter…"

Such an argument might occur because each of these three actually does create a different flower—the child a pink one, while the bee's perception constructs a red striped one and the cat's a dull gray flower! Bees see more colors than do humans, while cats have scarcely any color in their world. Bees need to know the world of flowers in order to make their living, but flowers do not matter a bit to cats.

Bats can *see* in the dark by sonar, as do dolphins in murky waters, by bouncing sound waves off objects in their environment. Dolphins and dogs create a good part of their worldviews from sounds we humans cannot perceive; many mammals live in a world made more of smells than of sights. Birds and insects sense the patterns of magnetic fields in the atmosphere.

Each species has its own system of senses that bring patterns of stimulation from the environment to its brain as it explores and does things to its world. These patterns coming in from outside merge with the existing inside patterns to create perceived images. The worldviews our minds are capable of projecting 'out there' fool us into thinking we see things the way they really are.

The interesting thing about considering the projected worldviews of different species is how clear it becomes that none has a truer picture of the world than does any other. A worldview made of smell patterns—with all their attached meaning for that species or individual—is no less true than one of light and sound patterns. Each species has evolved a way of constructing its world and then its worldview—a way that best helps it get along in that world. Each uses only part of all the information, or patterns, available to all Earthlife species collectively—the part it most needs to survive in health and do its part in the greater dance of life.

Only we humans know that all these different worldviews exist. We can record and measure light waves beyond those we see, sound waves beyond those we hear, chemical smells beyond those we smell, magnetism we do not feel. We alone can understand that our perception of the world involves only a small part of all the information available and do what we can to expand our range of information. We have figured out how to peek in on the worldviews of other species by using instruments that show the way they perceive or sense in their world. With sonar, we can see as bats and dolphins do; with microscopes we can look in on the world of microbes; and so on.

Surely this, too, makes us a special brain experiment—the only species of players able to understand something of what all the others are doing in the play. We humans have, in fact, a strange ability that no other species has, as far as we know—we are able to make ourselves the audience of the very play in which we are acting.

While there are always communications going on within and between species holons in holarchies, even among the cells of our body holarchies, no other species investigates the others as we do, trying to figure out what they are like and what they are up to. No other creatures think of themselves as *observers* of the whole world, indeed the whole universe—all the others simply participate in co-creating it. Nor did we ourselves think this way when we first became human. In fact, it is likely that we played human roles in the play of Gaian creation long before we could stand apart from the world in our own minds to see ourselves and others as players. What was it, then, that changed our worldview to the perspective of audience?

O   O   O

In bacteria and protists, there is a pretty direct link between stimulation from the environment and behavior in response to it. Their sensory parts are directly connected to the parts that behave by contracting, rowing away with cilia, or engulfing a food particle. In multicelled organisms, the stimulus may occur many cells away from the moving, behaving parts, so communications systems evolved—chemical and hydraulic systems in protists, funguses, and plants; ever more elaborate nervous systems in animals. But the more complex the nervous system became, the more it developed its own patterns to come between the incoming sensory patterns and the outgoing behavioral patterns. The connections, that is, are no longer direct, and the creatures' worldviews are determined as much or more by their nervous systems and life histories than by the new patterns actually coming in from outside.

In social species something else comes into play between the senses and behavior—the whole history of interactions among socially related individuals. There is, in a sense, a social brain or mind organized and shaped by social interactions and language over time, incorporated into the brain and behavior of individuals as they learn to live in society.

Language has played an enormously important role in the building of human societies and cultures. The human mind itself is largely a product of our social language community. Language is without doubt at the heart of our humanity. Written language may have been the invention that changed our mental images of ourselves and our world more than any other.

It was very likely writing that changed the way we saw ourselves in relation to the world in which we live—that permitted us to be observers of as well as participants in life's play. Before the development of writing, only pictures let people think of themselves as separate from their world—as observers, or knowers, of it. Today, we can see the history of the world in a book, or events in another part of the world in a film or video—the latest technologies for keeping records of ourselves and our world apart from direct or directly reported experience of it.

Before writing, language was not a *thing* in itself. Talking was simply a skill like walking. Nor could people imagine knowledge being passed on through language in any way other than through direct learning from another living person. Neither poetry nor law nor any other body of knowledge could exist without a live human knower of it before words could be carved in rock, inscribed on a clay tablet, or written on papyrus. Writing made us observers of our own play and gave us a way to store information and pass it on unaltered to our own and all future generations.

It's hard for us to imagine what it must have been like not to have separated ourselves from the world as observers of it, not to think of ourselves as separate from our knowledge, not to think of our languages as languages and our minds as minds, or our world as something to know about in our minds. Yet all of us were like this as small children before we were taught to read and to think about the world. In this sense, human infancy even today repeats something of the infancy of our species.

Before we invented writing, the script of Gaian creation existed only in consciousness and was stored only in geological records, in DNA, in the development of embryos, in nervous systems, to some extent in human minds constructed by language. Through writing we began to separate our knowledge and ideas about the dance from the dance itself—in a sense, to separate the script from its playing out.

But remember that the reality we humans see as ourselves and our relationship to the world around us is our own creation, our own conscious imagery. Our worldviews are rich in the images that language makes possible—an ever increasing wealth of linguistic portrayals of our human interactions with one another and with the rest of nature. These we accumulate in our beloved and useful linear time and over cultures through written records.

Even our pictorial art is influenced by the way the artist's mind talks about what it sees. Try imagining something without imagining accompanying verbal labels and ideas. The flower we saw earlier through a child's eyes as quite simply pink and pretty has become, as we say, all things to all people—a tasty morsel for a farmboy to feed his donkey, a symbol of beauty and a token of love to a lover, a source of perfume to a maker of cosmetics, a model of reproduction to a biology teacher, a solar energy plant to a physicist, a warning of fading youth to one who is advancing in age. Any one of us can change these *pre*ceptions into other *per*ceptions as often as we like.

The meaning we give things comes from the context in which we see them, and we supply this context not from the sense impressions we receive from our world but from the patterns ever evolving inside our nervous systems—patterns which reveal themselves as that richly complex self-organizing mind which ever composes and recomposes itself through individual and cultural experience.

It is our human heritage to continually work at making conscious, thoughtful sense of all these patterns. We embed all new information into existing brain-mind patterns, put these patterns into categories or contexts of meaning, add to them, change them, rearrange, distort, and enrich them until they make sense to us as part of our overall worldview.

O   O   O

We have wanted and needed to make sense of the world for as long as we have had our human type of consciousness, and we can do this only by using our free minds to create meaningful worldviews. Yet just because our brain-minds are so free, each *individual* human can see and understand the world as differently as can various other whole species. If all of us saw and understood the world in the same way, without being told anything about it by others, we wouldn't *have to* try to make sense of it, or try to teach one another just what kind of sense we think it makes. Everyone would just *know* how things were and what they meant. We would be like frogs, all of which see dark specks in just the same way and know just what they mean and what to do about them.

Human worldviews must be created through the personal experience of living in the world. At the same time, all of us must fit our personal experience into a worldview given to us by others. For if our own experience does not fit into the culturally approved way our fellows see the world, we will be thought quite mad. In fact, people with worldviews completely different from ones we call *normal* are commonly considered insane or profoundly handicapped.

Only by agreeing with one another on what the world is all about—on how to make sense of it—can we have human societies or cultures. Most of our individual worldviews actually come from our culture—from family, friends, schools, books, television, and so on—though all of us add our own special touches through personal experience and ideas. Our creation of worldviews is thus yet another way in which our brains are an experiment in freedom. While most other species have evolved their special way of seeing the world as they have evolved their behavior—building it into their body plans—we humans are free to improvise both our worldviews and our behavior.

When we look at human history to see what a people's worldview was in a different time and a different place, we see that worldviews have evolved along with the visible aspects of culture, and that there is a very powerful relationship between the worldviews that people hold and the kind of society they construct—an inseparable relationship, that is, between the way people believe their world *is* and the things they *do* to one another and that world. In practice, our worldview is our script for the play of life, assigning each of us our role within it.

The most important kinds of worldviews we humans have created are religious and scientific worldviews. In religious worldviews, a goddess or a god—or both or many deities—create(s) the world and then continue(s) to rule or look out for it in some meaningful, purposeful way. In the western scientific worldview up to the present, the world happens accidentally and runs mechanically without purpose. As we have seen, however, this is changing rapidly to bring western and eastern scientific worldviews together in the belief that the world is a self-creative manifestation of an underlying conscious source that may or may not be purposeful.

In both kinds of worldview, the human task is understanding how the world is *ordered*—by what god-given or natural laws it works, or what else gives it meaningful or at least coherent pattern. The way to this understanding, however, is very different in the two worldviews. In religious worldviews, the order is learned from certain people, such as priestesses or priests to whom it has been revealed, or from sacred writings such people or their followers have recorded. In scientific worldviews, the order or pattern of the world is learned from scientists who look for it in nature, make theories about it, and do experiments to test their theories.

Theories are no more, no less, than well thought out ideas or models of what various aspects of the world seem to be and how they work. Scientific theories, then, are ordered worldviews that can be tested against predictions we make from them, though we must expect them to change as we ourselves change and as we gain new knowledge. Until recently, religions had worldviews that were not to be questioned, but with new historical information some religious worldviews *are* now changing to come toward harmony with scientific worldviews from their own side.

Religion and science thus give us our cultural worldview—our image of the natural world and our relationship to it. Our worldview also includes our ideas of what our relations to one another are, or should be, so it includes political, ethical, artistic and other cultural images and ideas.

Until the last half century before the new millennium, it did not occur to people that they could have anything to do with creating their worldview. All through history, people thought the way they saw the world was the way the world really *was*—in other words, they saw *their* worldview as the *true* worldview and all others as mistaken and therefore false. Many wars, both hot and cold, were caused by disagreements between people who believed in a particular religious or political worldview and people who didn't believe in it—who had a different worldview that suggested different kinds of behavior and social structures. For example, the Christian Crusades against Muslims in the Middle Ages, the democratic revolutions against rulers in the past few centuries, and the more recent communist revolutions, were all of that sort.

People are very reluctant to change their worldviews, because their worldviews hold everything together—as we said, they make sense of the world. To change a worldview is to lose that sense, and so worldviews have usually changed only by force—when people with one worldview conquered those with another, as in the ancient conquests of Goddess-worshipping societies by God-worshipping tribes. In some cases, they have changed by persuasion, as when missionaries or scientists persuade people to adopt a new worldview in place of their old one.

Perhaps the most important discovery of modern science is that *there can be no single true and complete worldview.* Like all species, we have only partial information about the world, and our information changes as our knowledge increases, as our inventions become more sophisticated, and as we and other species actually change our world. We change the world even while we are looking at it, for we are never only observers—we are co-creative players in the play.

Most of what scientists do is try out—test by experiment—different parts of a scientific worldview, to see where it works and where it needs changing. Archaeologists and historians, along with biologists and physicists, conduct scientific searches, seeking experimental ways of testing their theories. It makes good sense to keep improving our worldviews as we gain new knowledge, to be sure they are reliable maps to the future we want.

While there is no scientific test for the truth of a worldview, there is scientific and practical evidence showing how well we get along in the world when we use a particular worldview as a map to guide our experience and behavior. We can test our worldview to see whether it guides us to as healthy an existence for ourselves and the larger life system of which we are part as do the worldviews that are programmed into other species. We will say more on this in later chapters. For now let us note that for the first time in our history, we as individuals are consciously evolving our worldviews by thinking about, questioning, and testing them, rather than just letting the events of history and a few powerful individuals dictate them to us.

O   O   O

Part of our evolving scientific worldview, as we said, is recognizing the validity and importance of other species' worldviews, expanding our own by incorporating theirs. It is equally important for us to recognize the validity and importance of different *human* worldviews, expanding our own in so doing. Every human culture that has its own language and customs has ways of seeing the world that are unique, though any human individual can learn any human culture and language. We can thus communicate across languages and share cultural experience in ways that enrich us all.

Today's dominant scientific worldview evolved in European languages with common roots and close relationships. These languages happen to be structured in a way that forces us, in talking or writing about our world, to think and speak of it in terms of 'thing-nouns' and 'action-verbs.' This language structure—taken for granted in English and other Indo-European languages—gives us a worldview, as soon as we begin speaking as children, in which we actually *see* the world as made of separate things that stay still (nouns) or move or are moved in relation to one another (verbs). The reasoning, the logic, and the mathematics of scientists are all based on this way of dividing up the world.

It can come as a great surprise to us that all people do not see the world in this way, and that this is related to language structures. Some human languages do not make our kind of distinction between nouns and verbs. Rather, the world is seen through certain languages as a pattern of interwoven processes in time, not as a pattern of separate things in space. Speakers of Hopi or Nootka, for example—both North American indigenous languages—cannot imagine things without their motion, change, aging, or other aspects of coming into and out of being. Instead of saying, for example, "The light shines," or "The water falls," they have single-word expressions that do not separate the light from its shining or the water from its falling. In such *process* languages, people do not think of time as made up of a series of 'things' called seconds, minutes, and hours. They see time as the change in things, which is more the way physicists now understand time.

These are only a very few examples of the way in which a language can determine how we see our world, yet they are enough to make us think about what the scientific worldview might be like if it had been developed in a very different set of languages. Einstein once suggested to Benjamin Lee Whorf, who studied and wrote about these language differences, that it might be easier to describe the discoveries of modern physics in the Hopi language than in English. In Hopi we would not face the contradictions of a world made at once of particle-things and wave-actions, of matter-things and energy-actions, never having separated things from actions in the first place.

Process-language cultures are better suited to organic than mechanical worldviews. Perhaps such cultures did not develop mechanical technologies because machinery is necessarily conceived and built as the interactions of separate and, insofar as possible, unchanging parts. As it happened, science developed most fully in European-language cultures along with machinery, becoming closely associated with machinery, as we will see in greater detail shortly.

Anthropology and linguistics, the sciences of human cultures and languages, are relatively new parts of science as a whole, but they have taught us that human experience is very varied and rich. They have made us realize that the scientific worldview, which was developed mainly in industrializing western countries, is based on and represents only a limited part of human experience. Many scientists, especially physicists and physicians, have begun to use ideas and practices from eastern worldviews to enrich their western science, while western science has become an important part of eastern life, especially in Japan and China.

Unfortunately, just as we dominant humans have worked at killing off other intelligent species, so have our dominant technological cultures worked at eradicating non-technological human cultures. Every culture and language lost seriously diminishes the variety and richness of human experience. This variety and richness is as essential to our cultural health as is genetic variety to the health of any species, and species variety to the health of any natural ecosystem. Our human mania for reducing variety to create monocultures is not expressed only in our agriculture and animal husbandry, but in our own cultures as well—and it is equally destructive in all cases.

Keeping in mind this perspective on our still-evolving scientific and cultural worldview—our continuing effort to understand what the play of life is all about—let's now look at its historical roots, its evolution within the cultural traditions of our species. Let's look, in other words, at the scripts people have written for themselves as they played out the historical steps leading us to our present conception of the play.

13

Worldviews from the Pleistocene to Plato

The earliest worldviews we know anything about date back to the *Pleistocene* epoch of ice ages—to what we call the *Paleolithic*, or Old Stone Age. In Europe and Asia, these sometimes cave-dwelling cultures were artistic and commonly symbolized nature as a great mother, a fruitful goddess who gave them life and all that was needed to sustain them. Seeing nature as the great Mother Goddess would have made sense of human experience by explaining nature's gifts of food and the birth of creatures, bright sunny days, like the goddess's good moods, and angry storms or droughts, like her bad moods. Nature seemed to love and to rage at her human children, giving them reason to celebrate her life giving and nurturing, as well as to love, fear, and respect her.

We can safely assume that paleolithic peoples took all nature to be as alive as they themselves were, and that they felt themselves to be part of, or the children of, this great mother. Even today, peoples who live in natural settings without changing them significantly tend *not* to divide nature into living and nonliving parts as our dominant culture does. Their only concept of non-living is of something dead that has been alive.

Later Stone Age and Bronze Age peoples must have called the Mother Goddess by many names in many places—only the later names, surviving into written language, were recorded. Just a few that *were* recorded are Nammu, Utu, Inanna, Ishtar, Iahu, Astarte, Ma, Kali, Isis, Gaia and Matrona. Iahu, meaning exalted dove, was the Great Goddess' name in Sumeria—the first great urban human culture—and was apparently later turned into the masculine Je-ho-vah. In ancient Greece they called her Gaia, as well as Eurynome and Demeter and Pandora, meaning 'giver of all gifts.'

The Mother Goddess of these ancient religions was surely not conceived of as outside of nature. She was not the external creator of nature but the creative force of nature itself. In modern religious terms, *Mother God immanent.* Nature was felt as the creative-destructive dance of life and death. Various components or forces of nature—Sun and Moon, winds and seas, mountains and rivers, animals important to people—were assigned to members of the goddess's 'holy family' or seen in other supportive deity roles.

The image of nature as a providing mother and the worship of this Great Goddess very likely influenced the development of Stone Age societies as agricultural 'households.' Archaeologists James Mellaart and Marija Gimbutas, as well as the archaeological scholars Merlin Stone and Riane Eisler, have given us an image of such early civilizations, as exemplified by the well-preserved Neolithic town of Catal Huyuk in Turkey. People of such societies provided for themselves and one another by raising crops and keeping tamed animals.

Most striking in such well-planned and managed agricultural societies, with their large towns, agricultural technology, beautiful wall paintings, decorated pottery, sculpture and metal arts, is that unlike later cultures they show no evidence of fortification, warfare, conquest, slavery, or significant social inequality, judging by house size, burial customs, and so forth. This is taken to mean that men and women worked in partnership, and there is evidence at Catal Huyuk that those in need were provided for from public stores of food or from the goddess's temple gardens.

Such ancient societies seem to have practiced the kind of peaceful life with all people's needs met that our modern societies are still far from bringing about. The remains of cultures throughout the Middle East, North Africa, and Europe, including pre-Minoan and Minoan Crete, show highly advanced societies, in which, as historian Riane Eisler puts it, "linking, not ranking" predominated.

The extent to which these societies were designed and managed by women will probably be debated among archaeologists for some time, but there are indications that women's roles were at least as important as men's. Mellaart found the 'holy family' at Catal Huyuk represented in order of importance as mother, daughter, son, and father, and a similar order was suggested in households by the sleeping platforms, that of the woman being more fixed and prominent than that of the man. There is no intrinsic reason to doubt that women, as the human representatives of the goddess, were accorded the social status that men gained later as human representatives of the god, but these early societies show no indication that men were oppressed by women; on the contrary, they indicate partnership.

If women did have the authority to make social rules, we might expect those rules to have been based on partnership for the simple reason that women give birth to and raise both girl and boy babies without considering the one better than the other, *if* the mothers are permitted to act on their natural feelings. The preferential treatment of boy children in some later cultures came about when men made the rules and set the cultural patterns. Creation stories of ancient societies often told of man and woman having been created together, as they were in the original Hebrew-Christian *Genesis* before it was rewritten to have Eve created from Adam's rib.

On the contrary, the hunter or nomadic Father God worshipers who invaded and conquered these societies were apparently not so peaceful and egalitarian. They were apparently headed by men who were experienced in the use of weapons. Perhaps they were driven to violent competition by their harsh environment and had come to worship lightning-bolt-wielding and thundering sky-gods in fear. After all, they were relatively unsheltered in open spaces, vulnerable to storms as well as to the marauding attacks of other, similar tribes.

When these conquerors invaded and stayed to rule a settled society where they found life good, they changed not only the social structure and rule but the society's worldview as well. Often they turned the Mother Goddess into the wife or daughter of their chief god and joined lesser gods and goddesses from both religions into a single *pantheon,* meaning 'all gods' religion.

Sometimes they got rid of the goddess altogether by making up stories in which the god was great and the goddess was only a disobedient mortal woman who was forever making trouble. Pandora was so demoted. Her name still means 'giver of all gifts,' but in the story we hear about her she brings only troubles into the world by disobeying the Father God. Similarly, the Hebrews, whose difficult wandering existence in the desert had somehow led them to believe in a stern Father God, turned the Mother Goddess, along with her symbols—the serpent of wisdom and the tree of knowledge—into Eve, another mortal woman who brought trouble into the world by questioning male authority and disobeying God.

Later, when Christianity replaced the pagan religions, old male deities were also contemptuously dethroned. The Celtic Sun god Lugh, for example, first became Lucifer, angel of light, and then was cast from heaven in medieval times to become Lucifer, or Satan, the symbol of evil.

All in all, the historical record tells us that when some men acquired the kind of power that accrues to those who have weapons and wealth, they formed a worldview based on a belief in their own superiority. They projected their self-image into an authoritarian and violent male god, thus justifying the domination of women, who came to be seen as the property of males, to be safeguarded and bartered. Nowhere is this more graphically recorded than in the Hebrew-Christian Old Testament Bible, and even the fabled Golden Age is tarnished by a similar treatment of women. Such male rulers extended the idea of superiority and the practice of violence into their affairs with one another as well—making war upon each other, dethroning the deities of the conquered, making warriors their heroes, taking slaves and building class-structured societies.

Another important aspect of the shift from a worldview based on partnership to one based on domination, as god-worship replaced goddess-worship all over the civilized world, was the idea that nature was separate from both gods and people—that it had been created and was ruled over by one God who was *external* to nature. Nature, as God's creation, was then seen as a gift given to His people to use and exploit for their own ends—as in the biblical "to have dominion over." The Old Testament testifies to a jealous and unmerciful God who urges man to make war on and destroy all non-believers and other enemies, and to subjugate women. His story became history.

And so, at the same time—a few thousand years before the Christian era—humanity seems to have undergone the two greatest changes in history since the advent of agriculture. One was the shift from the worldview and culture of partnership to that of domination—from the worship of life giving to the worship of life-*taking*, as Eisler puts it. The other change was the shift from a worldview in which people and their deities were part of nature's own improvised dance, continually self-creating from within, to a worldview in which men and their gods stood outside and above nature, in which men claimed the god-given right to exploit women and all the rest of the natural world.

All this, of course, is a sweeping simplification of history for the sake of seeing broad patterns. The role of women as the glue that holds society together cannot be doubted even today, but the partnership status, the equal valuation of their work and their arts, has never been regained to this day.

One of the latest goddess cultures to survive was that of Crete, known to the ancient Egyptians as the Keftiu, to us as Minoan. A peaceful agricultural people we mentioned earlier, the Minoans left us exquisite art in admiration and praise of nature. Nature goddess worship was evident in other parts of Greece, too, and lasted in some form until classical times—even Plato being initiated into the Eleusinian mysteries of Demeter. But the sequence of Greek myth, as Robert Graves pointed out, shows the gradual destruction of equality and goddess-worship in favor of patriarchal rule and god-worship.

O   O   O

Only the earliest known Greek philosophers were deeply influenced by ancient ideas of nature's self-creation, as were early philosophers in other lands. By the sixth century before Christ, most of the civilized world had been organized into large kingdoms or empires with generally patriarchal religious worldviews and strict laws for keeping order.

And yet, sixth century BC thinkers such as Lao-tse and Confucius in China, Vedist Hindus and Gautama Buddha in India, Zoroaster in Persia, and Thales, Anaximander, and Heraclitus in Milesian Greece (now Turkey) all came to very much the same idea about how nature works. In carefully observing and thinking about nature, they all saw it as alive and forever changing from within, whether or not it was symbolized by a pantheon of gods and goddesses. They saw nature as striving to create its own balance and order through an endless dance of opposing forces or principles such as male and female, light and dark, hot and cold, inward and outward, storm and calm, creation and destruction. In this dance, opposites clashed or simply got out of balance so that things grew, say, too cold or too stormy or otherwise disorderly. Yet somehow new forms and patterns created themselves to bring about new balance and harmony.

Even though they could not talk to one another, these great thinkers all over the civilized world of the sixth century B.C. somehow agreed that nature's constant movement was away from disorder and toward balanced order—what we now call "order out of chaos." This balance, or harmony, they believed, must ever be re-created from imbalance or discord, very much as it is in human affairs. This did not surprise them, because they all saw humans as part of nature.

In Greece such thinkers formed the first scientific worldviews by trying to understand and explain the world in terms of what they could see in nature. Poets, meanwhile, continued to spread the Greek worldview of Gaian creation and the Olympian pantheon of gods and goddesses who ruled the world, as did Homer in the Iliad and Odyssey, which had, by then, been written down.

The new scientific thinkers came to be called *physicists*—as the Greek word for nature was *physis*—or *philosophers*—from the Greek words *philos*, meaning lover or friend, and *sophia*, meaning wisdom. The wisdom they loved and sought after was an under-standing of how nature works, because they believed that by under-standing the natural order they would come to understand how to order human life, both personal and social, more wisely.

In the myth of Gaia, recall, the goddess comes out of *chaos* to transform her body into the Earth by her dance. Originally, chaos meant nothingness; later it came to mean anything that seemed to have no pattern, that was completely lawless or disorderly. The oppo-site of chaos was order, which was called *cosmos*, still today the Greek word for world. The world, in other words, is the pattern of things.

The eastern Greek Milesian philosophers agreed that the natu-ral world is orderly—that it has a pattern that can be discovered, described, and understood by human beings. As scientists, they saw orderly rhythms, balance and harmony in the patterns of stars and planets, the cycles of seasons, the beautiful forms of plants and animals. Disorder, or chaos, disturbed this order, but it seemed that wherever disorder arose, order was quickly restored. Birds and worms ate up dead animals; old leaves disap-peared into rich new soil; rain made droopy plants grow healthy and flower; new forests grew from burned ones. Nature kept making orderly patterns out of chaotic disorder. And what was so interesting about all this was that everything in nature played its part without being told what to do. This observation came to play a very important role in Greek politics before long.

Plants took form, growing from seeds, then rotted back into soil, losing their form. Older animals died as young were born in an endless chain of life. One creature ate another to live itself. Nature was one great intertwined pattern in which, as the philosopher Anaximander said, *"Everything taking form in nature incurs a debt, which must be paid by dissolving again so that other things may form."* He saw this as a kind of justice—each thing, or creature, in nature borrowing from nature's supplies, then paying them back. Rivers dried up while new rivers formed elsewhere. Clouds formed, dissolved in rain, and left clear skies, which later formed new clouds. Fires and storms created chaos, yet from the chaos of destruction new life and new order always arose. Everything that took on its own form later gave way to other newer forms.

Anaximander's teacher, Thales, thought all things in nature had formed from water, and had water as their essence. Anaximander himself, seeing the fossils of sea creatures on land, thought about the great changes in geology and in life forms that must have happened over time. He came to believe that living creatures first formed in the seas, later came out onto dry land and shed their shells. Humans, he reasoned, must have been born from earlier animals, since the first human babies could not have taken care of themselves. As far as we know, he was the first scientist to see a pattern of evolution by actually observing nature. The way in which nature was understood by Anaximander, his teacher Thales, and *his* pupil Heraclitus—all Milesian Greeks—was very much the way scientists are beginning to understand it again now—as a great dance of life in which all natural things are connected and constantly improvise their steps as they move toward balance and harmony.

These philosophic ideas seem to have echoed a distant time when people actually lived in democratic balance and harmony within the larger context of nature. Some memory of these times was recalled by the poet Hesiod, around the same time as Homer, when he wrote (as quoted in Eisler's *The Chalice and the Blade*) of a former goddess culture he called the Golden Race: *"All good things were theirs. The fruitful Earth poured forth her fruits unbidden in boundless plenty. In peaceful ease they kept their lands with good abundance, rich in flocks and dear to the immortals."* This race, Hesiod continues to tell, was later conquered by a lesser race of silver and then *by "a race of bronze, dreadful and mighty, sprung from shafts of ash,"* bringing war. *"The all lamented sinful works of Ares [God of war] were their chief care."*

It is interesting to note that in the original myth of Gaian creation, the Olympic gods were born of the giant Titans, who were the first children of Gaia and Ouranos, the sky whom she created. Were the Titans perhaps a symbolic memory of the powerful patriarchal tribes—the Achaeans who destroyed the goddess's rule at Delphi and put Apollo in her place, the Aeolians and Ionians who overran Greece in later waves?

In the sixth century B.C., the ruler of Athens, Solon, put into practice philosophical ideas of natural balance and perhaps also the mythologic-historic memories of greater equality. Trying to create some semblance of democracy, he divided land more equally among people and made laws to ensure greater justice and to give citizens more say in the decisions of their society.

The playwright Aeschylus, whom we mentioned earlier, also used these ideas in his dramas, giving his heroes and heroines the task of balancing the scales of justice in working out their personal lives within the larger framework of family, society, and all nature. But we also see in Aeschylus how this process of justice was undermined in the shift from a goddess culture to a god culture by the devaluing of women.

At the beginning of Aeschylus' famous trilogy about the Mycenian house of Atreus, Queen Clytemnestra's murder of her husband Agamemnon is personally justified by his previous sacrifice of their daughter, Iphigenia, and socially justified by the queen's status as head of her clan, with the responsibility for avenging bloodshed. At the end of the trilogy, her son Orestes is tried by the new court of Athens for murdering his mother in revenge and acquitted on the grounds that he has not shed kindred blood. Athena, no longer the ancient nature goddess, but the warrior child of Zeus, sprung to life from his ear, presides over the trial and casts the deciding vote. As Apollo explains, *"The mother is no parent of that which is called her child,"* but *"only nurse of the new planted seed that grows."*

This, then, was the beginning of the Golden Age of Greece, which all the world remembers for its beautifully harmonious temples and theaters, for its sculptures and Olympic Games, and for its experiment in male political democracy.

It is important to understand that when this limited democracy was formed, the Greeks had no concept of perfection in their worldview. Their traditional gods and goddesses were seen, like people themselves, as part of nature—imperfect, moody, and mischievous, often intentionally creating disorder, from which they then made order under the higher law of justice. Nor did the Milesian philosophers, the first scientists, see nature as perfect. They knew well that nature was never perfectly balanced or harmonious, but always struggling *toward* balance and harmony. Wherever it was won, it was soon followed by new imbalance that drove the dance forward in search of harmony.

If nature reached perfection, its evolution would come to a stop. If things fell back into complete chaos, creation would also cease. Nature's dynamic balance is always achieved somewhere between chaos and perfect harmonious order. There was certainly no reason in this Greek worldview to expect men or their society to be perfect. On the other hand, there was hope that neither men nor the society they were trying to balance would fall into complete disorder.

The experiment of the Greeks in trying to make order out of chaos by ruling themselves democratically instead of letting rulers tell them what to do was not an easy one. Would men be able to agree on how to balance their society and live harmoniously? Balance in society would mean allowing all male citizens to take equal part and responsibility in how things were run. Harmony would mean a love of the good life not only for oneself but for everyone else, too. Men would have to make choices that were good for themselves *and*, at the same time, good for all society.

O   O   O

While the Athenians struggled toward democracy for imperfect men in an imperfect world, the western Eleatic philosophers on the other side of Greece from the Milesians were forming a new kind of worldview. They had become fascinated with the human mind itself, and with a kind of perfect order it had created, a perfect order they had found not in nature but in the man-made language of mathematics.

The Greeks had used mathematics to measure the size and distance of things, to map the heavens and the Earth, to make calendars, and to predict events such as eclipses, as had other ancient peoples such as the Egyptians and Babylonians. These were practical uses of mathematics. But mathematics was a wonderful invention by itself because of its perfect order.

Arithmetic and geometry were languages of number and form built on rules of balance and harmony—but not the kind of natural harmony that falls into disorder and has to rebalance itself by creating new forms. Mathematical rules *kept* things in perfectly balanced order. Geometric figures—such as circles, spheres, cubes pyramids, octahedrons and others—were the most perfect things that humans had ever created. Furthermore, their perfection never changed, never fell into disorder.

The ancient Eleatic Greek philosophers, who lived in what is now southern Italy and Sicily, with Pythagoras as their first mentor, sought these perfect forms and harmonies in sacred geometry. They decided the Milesians must be wrong about a natural cosmos evolving through the ever changing, or *dynamic*, balance between chaos and order. The Eleatic worldview was of a cosmos created in the mathematical perfection of unchanging balance and harmony. They held that nature only *appeared* to be imperfect because people were blind to its underlying perfection.

In this worldview, the stars and planets were held on perfect invisible spheres that turned around the Earth, which was at their center. Stars were the effect of cosmic fire shining through holes in these spheres. Pythagoras' discovery that musical harmony obeys mathematical laws gave birth to the idea that the heavenly spheres created music in their turning.

This worldview permitted no evolutionary change—only a perfect turning around and around, a perfect repetition of the same cycles in the heavens, and of the same cycles of birth and death among the creatures of Earth. Some Hindu traditions maintain worldviews of ever-repeating cycles even today. For the Eleatics, everything had been just as it was now from the very beginning of the cosmos. In place of the uncountable opposites unbalancing and rebalancing themselves in the Milesian worldview, Empedocles proposed four unchanging elements that mixed in different proportions to form things. Parmenides even arrived at the idea that nothing in the cosmos or world moved at all—that the whole appearance of motion in the world was an illusion, a strange trick of human perception. And Zeno, whose mathematical puzzles still fascinate mathematicians today, 'proved' that the world had neither motion nor parts.

Parmenides and Zeno formed these ideas by using their minds to think about how things *must* be beyond their appearance. There cannot be anything that does not exist, Parmenides reasoned, so all the cosmos must be filled with things that do exist. And if it is completely filled with existing things, there can be nowhere for them to move. His pupil Zeno was even better at showing logically that nothing is as it seems to be. In using pure thought to map their world, these philosophers abandoned scientific observation of nature.

Other philosophers, even if they did not agree that there was no motion at all, did come to feel strongly that the human mind could understand nature better just by thinking about it than by observing it through their own senses. Democritus, for example, came to the idea that everything in nature is made of invisible tiny hard bits he called *atoms*—from *atoma* meaning individual—and that the motion of atoms combined them into the things we see. These atoms were eternal and perfect, as they never changed and could not be destroyed.

More and more the philosophers felt that senses such as sight, hearing, and touch, through which we get our everyday experience of the world, were less trustworthy than the reasoning mind. The reasoning mind, which had invented arithmetic and geometry— rules for ordering numbers and shapes—now invented *logical* rules for ordering human thought, for making it as much like the perfection of mathematics as possible. Thoughts and ideas ordered by logic could be written down, compared with other philosophies, polished, and perfected. Such a philosophy could exist, like arithmetic or geometry, as a thing to be known in itself and passed on to future generations.

Of all the amazing Greek philosophers exploring these various worldviews, the one who was most fascinated by the way the human mind formed its ideas was the Athenian Socrates. For all we know about Socrates, it is not often mentioned that he acknowledged the priestess Diotema as his teacher, nor that Pythagoras had acknowledged Themistoclea in the same way. Socrates was much less interested in what the natural world was and how it worked than in what the human mind was and how *it* could be made to work better than it did in most people. If ordinary people, not only philosophers, could get clear on just what they *meant* by the good life and good government, for instance, he felt they might figure out *how* to improve their lives and their government.

The limited democracy of Athens had become a kind of shouting match in which the cleverest speaker won the day, whether he really knew what he was talking about or not. Some men were simply looking out for themselves and seeking ways to gain power. If they argued loudly and well, others too lazy to think for themselves would vote their way. Many citizens balked at having to listen to one another's harangues and vote on them. Squabbles continued and men complained about their responsibilities as citizens; others took advantage of the confusion to try to bring back dictators. The playwright Aristophanes wrote some very funny plays about solving social problems in spite of the lying, cheating, lazy, yet clever Athenians. He also resorted, in several of his plays—*Lysistrata* and *Women in Parliament*—to the idea that women might solve the social problems of government and war better than men could.

Socrates—immortalized by his star pupil in *Plato's Dialogues*—was meanwhile spending all his time cornering people and pressing them with questions to help them think more clearly, and he was much loved and admired for this by his followers. But his criticism of the muddle democratic government had gotten into finally led to his trial and execution.

Socrates saw people as throwing their ideas together from anything at all that came into their heads—like builders trying to make a building without a plan, throwing together any materials they happened to find lying around. So he tried to teach them to decide just what they wanted to think about, how to recognize and throw out useless ideas, how to make muddled ones clear, and how to build the clear and useful ones into the understanding they sought—in other words, he taught them to think in orderly ways, to reason *logically*.

Plato was influenced by the Eleatic search for perfection and fascinated by the beautifully clear ideas or definitions Socrates was able to arrive at by reasoning. Thus he concluded that perfect ideas must really exist somewhere behind the muddled world we normally see. Not only, he reasoned, was the material world beyond the senses perfect— with ideal forms of chairs and trees and everything else in our sensory world—but so also was there a world of perfect ideas, such as justice, love, truth, and beauty. Our senses shackle us, said Plato, showing us merely the flickering shadows of a perfect world beyond—a world of light we can reach only with our minds. It is interesting to compare these ideas with those of eastern philosophers teaching meditation to reach a perfect state of merger in the source reality of loving oneness.

Plato's ideal world fit in with other philosophers' ideas of the cosmos as a construction of perfect spheres, a world built from perfect atoms. All the older worldviews seemed childish in comparison with this elegant new one 'discovered' by reasoning minds. What could a perfect world, or cosmos, have to do with an unpredictable, moody goddess or with disorderly gods and goddesses who lied and tricked one another to get out of the messes they made? Poets began making fun of the old religion while philosophers began thinking of a new one.

A perfect world had to be the work of one perfect creator, Plato reasoned, a God, who existed apart from this shadow world, who created a perfect and unchanging world, *a God who was always doing geometry!*

The old philosophy of nature as alive and creative in its imperfection was replaced by belief in the perfect and rather mechanical creation of a single, though yet unknown, God. Perhaps this new worldview was comforting to the Athenians at a time when they were having so much trouble working out democratic order. At least they could believe in a perfect world just beyond the mess they were stuck in.

14

Worldviews from Plato to the Present

Eventually, through an itinerant preacher—or storyteller, in Greek tradition—named Paul, a new religion came to Greece from the East. It was a religion that fit Plato's worldview very well. Not only did it explain the creation of the world by a perfect God; it also explained the disorderly ways of humans as disobedience to God, their Father. This gave people a new hope—that they could make themselves and their society more perfect by obeying God's law.

According to this Hebrew-Christian worldview, God had created the world only a few thousand years before, with all its different kinds of plants and animals, just as it is now and right at the center of the universe. This view was very much in keeping with Plato's brilliant student Aristotle's static view of nature, Anaximander never having been taken seriously on the subject of evolution.

The scenario is familiar—God created the world as a paradise for humans, setting the first two people into its perfection. They

were expelled after Eve disobeyed God's law by tempting Adam to
join her in the sin of eating fruit from the tree of knowledge, thus
bringing disorder and strife into the world forever after. Still, paradise could be regained after death, in a heavenly world, if people
became perfect in God's eyes again, and this perfection could be
accomplished by seeking forgiveness and obeying God's law.

In the ever expanding thrust of empires, Rome conquered
Greece. Though Jesus himself had preached equality for all, including women, and had been opposed to any kind of dominance, the
holy texts were rewritten to suit the priesthood of the new church,
after the earliest Roman Christians had been tortured and killed.
By the time the Byzantine Empire split off from the Roman
Empire, both empires, with slight differences had officially
adopted the church's revised Christian worldview, adapted by
priests to their hierarchical societies.

The rich heritage of ancient Greek scientific discoveries—that
the Earth moved around the Sun, that nature was alive and evolving, that humans were descended from simpler creatures, that
many of their ills were curable by medicines and surgery—was
destroyed, forgotten, or denied as the new worldview took over.
The great library at Alexandria was repeatedly sacked and burned
by Romans, Christians, and Muslims; nearly a million book-scrolls
of human knowledge and culture were lost.

In Europe, for well over a thousand years, all ideas—scientific
and other—that did not reflect the Christian worldview were considered heresies and outlawed wherever possible. Brutal Crusades
marched outside of Europe to attack Muslims as non-believer infidels; women were burned to death in droves as witches all over
Europe—for practicing the older nature religions persisting among
ordinary people and healing with natural medicines. Christian

priests, who explained God's law to people and enforced it, were the rulers of all Europe. Even kings bowed down to the highest priest of all, the pope of the Holy Roman Empire. Waves of plagues were seen as God's wrath released against his disobedient children, and the church grew ever stronger for its presumed power to grant salvation.

Many of Plato's ideas about education and politics in a perfect society were put into practice during the Christian era. Plato had written, for example, that a perfect society should be ruled by the most educated of citizens, people from all walks of life who lived simply, without personal possessions. Though Plato had not advocated the exclusion of women from education and rule, the Christian monks and priests, who wielded so much influence in Europe, otherwise met Plato's requirements. The highest-ranking clerics did indulge themselves, however, in affluent comforts and fine robes. The idea of heaven and hell as places where people would go after death to be rewarded or punished for their Earthly behavior also came from Plato's writings. And the works of Aristotle taught Christian Europe formal logic and the pursuit of virtue—originally meaning excellence, but later coming to mean obedience.

For more than a thousand years the Christian Europeans educated only boys and men, teaching them the invented mechanical languages of mathematics and logic in the dead language of Latin. A dead language is one no longer in common use, no longer imbibed by children at their mother's knee, no longer changing, but frozen in time. This education is important to remember when, in the next chapter, we consider how the differences between natural living languages and artificial nonliving languages affect worldviews.

For now, let us note that the perfected mechanical languages of mathematics and logic played a central role in the rebirth of science in Europe. This rebirth was of course part of the larger rebirth—or *Renaissance*—of human curiosity and culture that began about five hundred years ago.

Through trade with the East, opened up by Crusades, some Europeans, such as the Medici family in Italy, became very wealthy and very worldly. To have beautiful and interesting things around them, they hired architects, artists, and scientists to create splendid new works and to seek new knowledge. Of note, this expansion of urban life and culture also led to vast deforestation all over Europe, as ancient Rome had deforested northern Africa to build its ships and cities, as well as to open land for growing its grain.

Traders and refugees brought copies of surviving ancient scientific writings from Constantinople, from Arab lands, from Moorish Spain, from wherever they had been preserved, studied, and further developed, as they were especially in Muslim mathematics, astronomy, alchemy, and medicine. These manuscripts, including many Greek works salvaged from Alexandria and preserved by Muslims, reawakened interest in questions of planetary movement through the skies and of the location and nature of the Earth, with all its living plants and animals.

Giordano Bruno, the first philosopher-scientist to revive the ancient notion that the Earth moves around the Sun, was burned at the stake in the year 1600 by the Christian priests of the Holy Inquisition. But only ten years later Galileo Galilei built the first telescope and with its help showed that the Earth *does* revolve around the Sun and so cannot be at the center of the universe. Neither Bruno nor Galileo ever meant to disprove the religious worldview, but only to *improve* it. Yet Galileo, too, was punished by the church. Narrowly escaping the stake, he was imprisoned and forbidden to teach.

It is important to remember that all the founding fathers of modern science were religious men eager to show the glory of God by giving people a better understanding of His wonderful creations. They imagined God very much as Plato had—as a geometer. Mathematics, Galileo said, was the language in which all nature was written. And so the most important task of reborn science was to discover the mathematical laws by which God had created the world.

O   O   O

Throughout the Middle Ages, the Renaissance, and even later during the Age of Enlightenment, as Carolyn Merchant documented in *The Death of Nature*, a belief in nature as alive, personal, and mysterious, persisted among ordinary people as well as in the tradition of *alchemy*. Yet as modern science evolved, it weeded out these ideas in favor of a belief in nature as an impersonal mechanism that had to be brought under human dominance by rational understanding and mathematical description. Let's see how this dramatic change in worldview occurred.

The ancient Greeks, especially Archimedes, had already begun to make mechanical models of how things in nature work. In fact, the words *mechanism* and *machine* come from the ancient Greek word for these models. Archimedes built actual machines that were very ingenious and successful—especially some rather amazing machines for fighting wars, such as huge cranes equipped with claws that could life ships out of water and smash them on ground. But he was so ashamed of these crude imitations of sacred geometrical designs built for practical mundane purposes that he never even wrote about them himself. Greek philosophers felt that physical machines were poor imitations of God-the-geometer's sacred works—such sacrilege that ancient Greece did not develop its technology.

The Renaissance, however, gave rebirth not only to art and science but to mechanical engineering as well. Scientists themselves began building mechanical models to help work out the geometric patterns of the heavens wheeling around the Earth. Later models showed the Earth and other planets wheeling around the Sun while Moons wheeled around planets, all against the greater background of wheeling stars. Each heavenly body they knew about was attached, in these models, to its own ring of metal in the great sphere of the universe. Some of the most elaborate models with the greatest number of rings looked a bit like huge sculptured balls of metallic yarn. All these rings wheeling about within one another suggested that the universe must be something like other Renaissance machinery—something like the clockworks in church towers, with various wheels turning other wheels in their mechanisms.

René Descartes—another founding father of modern science—invented a new mathematics along with a whole new framework for the religious-scientific worldview. In this view, God was not only a geometer, but also a Grand Engineer. Using mathematical laws, God had not only created cosmic clockworks, but had put into it endless smaller mechanical inventions such as plants and animals and people. Descartes insisted there was no essential difference between man-made machines such as clocks, grain mills, or jeweled golden wind-up birds that sang and the living mechanisms God had created. God's were more complex, but man could learn that complexity. This became the dominant worldview of all science.

As men were God's favorite mechanisms, Descartes explained, He attached to them inventive minds that worked quite like His own. Women, like animals, had no such minds, and were to be controlled by men along with the rest of mechanical nature. Since men's minds had been made to work like God's own, it was no surprise that men, too, were inventive engineers, putting their own mechanical robots into their own mechanical clocks high upon the church towers in honor of the Grand Engineer.

None of the scientists who accepted this worldview ever seemed to wonder if man had not projected his own mind, talents, and achievements into his image of God, rather than the other way around. They were convinced that, except for their own male minds, everything in nature was God's mechanical creation, made to be understood by men. Surely this was as strange a worldview as humans had ever held, yet it wove the religious worldview and the scientific worldview into one and gave scientists new visions of understanding and controlling all nature, as they were convinced God intended.

Imagine the excitement they felt—if everything in the whole world, even in the whole *universe*— was mechanical, then men who understood mechanics could understand how all nature worked by taking everything apart to see what made it tick! And sooner or later, surely, they would be able to make their mechanical birds as good as God's feathered ones.

Another founding father, Francis Bacon, who is credited with having developed the scientific method, wrote much about the coming Golden Age of Science, when man would understand and control all nature, creating his own mechanically perfect societies, which would be free of all human problems. Wasn't that what God intended for his favorite creatures? Bacon was a lawyer who attended many witch trials. He, like many men, identified mysterious nature with woman. Science, he said, would flourish when men grew up and stopped expecting her to unveil herself at their request, but instead hounded nature and tortured her secrets from her.

It is interesting to contrast Bacon's vision of a future Golden Age with ancient Hesiod's lament for the times of a past Golden Race—to reflect on what seems utopian in the context of different worldviews. For Hesiod, peace and bounty seemed to pour from nature itself, whereas for Bacon all good things were to be wrested from nature at any cost by dominating and controlling it.

O   O   O

The mechanical worldview suited the next few centuries very well, for nothing mattered more to the Europeans and their off-spring Americans than the machinery that was changing the whole human way of life. Scientists, living in a society that was becoming ever more mechanized, saw more and more mechanisms wherever they looked in nature. Geologists described geological mechanisms—how the Earth was put together, how the cycles of weather ground up rocks, and so on. Biologists spoke of the mechanisms of living things—how the parts of plants and animals and people were put together and how they worked. In time, doctors spoke of heart and lung pumps, of bone and muscle mechanisms; still later psychologists studied the machinery of the brain and social planners worked on the mechanisms of society.

Scientific discoveries of 'natural mechanisms' depended on the invention of new man-made mechanisms such as telescopes, compasses, thermometers, barometers, scales, clocks, and later more sophisticated devices, all of which made it possible for the scientists to detect and measure ever more parts of the natural world. Science also depended on the invention of new mathematics for modeling relationships among these measured parts of the world, *for only*

*measurable parts of the world could be studied by scientists with a
mechanical worldview.*

Let's look for just a moment at this role played by mathematics in
science. Mathematics itself is not a science—it is the art of making
complicated and beautifully balanced patterns from very simple
basic symbols and rules for combining them. Mathematicians can
keep finding new patterns to make from the basic symbols and rules
they have adopted, or they can change those symbols and rules to
develop a wholly different set of patterns.

In pure mathematics, the symbols have no real-world meaning—no one has ever found a 2 or a + or a > in nature. But scientists have found that when they assign a real-world meaning to the
symbols, some mathematical patterns turn out to be very useful as
models of the measurable aspects of nature they study. Many mathematical models are designed to be translated into physical mechanisms, as they were in the Renaissance models of the cosmos.

Consider that whenever we want to describe something previously unknown to us—when we want to understand it—we must
find something familiar to us with which we can compare it. The
known thing becomes a *metaphor*—literally a carrier—for understanding the unknown thing. This is because the brain can use
new incoming patterns of information only in the context of its
existing patterns.

Every day we use metaphors such as "This material is as soft as a
baby's skin and as blue as the sea," or "That man is watching me
like a hawk," or "The boss wants everything to run like clockwork." But how often do we reflect on the fact that all new understanding, even scientific understanding, comes about in this way?
The patterns or forms of machines and mathematics are human
inventions, and thus they are very familiar to their inventors. So,

when scientists use them as models—saying, for example, that hearts pump blood and plants pump water—we should be clear that these are simply metaphors that help us understand something about hearts, plants and their energy use.

The idea that reality is made up of only measurable things, and that their description is the only possible knowledge of reality, is called *positivism*. The tasks of positivist science have been seen as twofold—to discover what the parts of natural mechanisms are, and to see how the mechanisms work through the movement of these parts. In other words, scientists took things apart in order to see of what they were constructed as well as how they 'ticked.' This method of reducing things to their parts came to be known as the *reductionist method* of science.

In so reducing things to their parts, scientists showed us a fascinating inner world of things. Live bodies, for instance, were made of bones, skin, blood vessels, nerves, and other organs; each organ was made of tissues, each tissue of cells, each cell of chemicals. In fact, everything turned out to be made of chemicals, which in turn were made of molecules.

Molecules that were fixed in tight patterns so they could not move formed solid things, such as rock or wood. If they slipped and slid around each other, they formed liquids. If they floated about loosely, they formed gases. And molecules, it turned out, were made of atoms. At last scientists had the instruments and the mathematics to show that all natural mechanisms were really made of those tiniest and most indestructible building blocks called atoms—just as the ancient Greeks had said more than two thousand years before!

Only one problem—things did not prove to be quite so simple. Atoms turned out to be made of parts themselves, and their parts turned out to be anything *but* solid machine parts. Atomic physicists, just when they reached the very foundation of mechanical nature, discovered that nature is not so mechanical after all. But before we go on with this story, let's go back to look at the other part of what scientists were doing—seeing what the movement of parts was all about in natural mechanisms.

From their measurements and models scientists worked out mathematical laws of motion among the parts of nature's mechanisms. The more they studied motion in the universe, the more the universe seemed to move and change. Not only did the Earth no longer stand still at the center of perfect heavenly spheres, but it turned on its own wobbling axis and wheeled around its Sun in spirals, because the Sun itself wandered through space, dragging its planets with it as part of a galaxy moving on its own, and so on.

Geologists, digging into rock and studying landscapes, discovered that the Earth itself has changed a great deal over time. Biologists meanwhile grew curious about the fossils geologists uncovered. Earth seemed to contain her own record of plants and animals that had lived long ago, and the record indicated that they had changed a great deal, too.

How could the Earth be only a few thousand years old, as measured from the generations of people listed in the Bible from the creation of Adam to historically known kings? The geological record was proving it to be very much older, with different kinds of plants and animals at different times in its history. Had God created these different kinds of plants and animals at different times, rather than creating them all at once? Did He keep making them more complex with each wave of creation? Or had they somehow changed by themselves?

Once the idea of evolution, buried since Anaximander's time, emerged again, it quickly made a great deal of sense, and the whole scientific-religious worldview was turned upside down to fit it. Creation had been seen as a kind of ladder with God at the top. On the next rung down were the angels, then people, then the large animals, then the smaller ones, and on down to lowly worms and even smaller things, all Earth creatures having been created at once by God. In the new evolutionary view, the ladder began at the bottom with the most ancient, tiniest creatures, which changed over time, forming new rungs of ever larger creatures, climbing the ladder up to the rung of humans, who seemed to be at the very top— for scientists were beginning to doubt the existence of angels and God. Eventually, scientists gave up the whole concept of God and declared their separation from religion, a matter we will go into further later on.

When Darwin's theory of evolution through natural selection of the fittest in a great competition for limited resources became popular, the industrial age was well under way. In fact, the industrialists of Darwin's England were in just such competition for survival with one another, so they readily adopted the new evolutionary theory as part of their worldview. These ever wealthier industrialists were *not* so ready to believe the news that they were cousins to the apes, but the idea that they were the fittest creatures in all nature seemed to make up for it. They did not need to lose sleep over the poverty and toil they were forcing on their workers in factories and colonies, for their own riches and comforts were simply proof of their natural fitness. In fact, they took Darwin's theory as evidence that their way of life—industrial competition—was the most natural and the surest way of human progress. Did it not follow that a competitive capitalist society was the best possible social mechanism for producing the fittest humans through natural selection?

Not long after the theory of evolution became known, the Russian Revolution produced a new 'social mechanism' known as communism, which was heralded as being based on cooperation rather than on competition. Russian scientists rewrote the theory of evolution accordingly, to show that cooperation in nature produced more fit natural creatures than did competition!

O   O   O

Moving away from religion, science came closer to the politics of industrial man, who had wrested social and political power from the church. In both the capitalist and communist worlds, scientists were awarded the status of a secular priesthood with the mission of forming the cultural worldview—the story of how things came to be. In turn, they were supported by governments and rewarded for shaping worldviews consistent with the politics of their societies. Yet much as they argued the natural advantages of competition on one side and cooperation on the other, industrialism itself evolved and dictated a similar way of life on both sides for those who owned the means of production and those who worked for them. Industrialism, that is, shaped human habits to its needs—which have now shown themselves as greeds—making society itself into the great mechanism so wonderfully spoofed by Charlie Chaplin.

City transportation systems were built to get workers to and from factories, and education systems were designed to produce the workers. Schools trained children to be on time and to sit still for long hours without talking to their neighbors, doing what they were told even if it was boring, as they would have to in factories when they got older. It was as if children were raw materials put into the school machine and turned out as workers. The clocks and schedules of industrial workers replaced the Sun and the weather in telling people when to do what. Government systems got more complicated, more centralized, more organized, to manage society in ways that made industry and the trade of industrial products work smoothly.

Thus, families, schools, hospitals, governments, and other social institutions were run as efficiently as factory machines. The whole way of life became as mechanical as the scientific worldview, and new branches of science—economic science, political science, sociology—were created to design and build the machinery of society, to keep it well oiled and in good repair.

The idea of perfecting humanity, first stated in Plato's worldview, was held throughout the Christian Middle Ages and the Renaissance, and that same idea fired the imaginations of the founding fathers of modern science and industry. Now it was put to its greatest test—the modern industrial age was to bring the solution to all problems at last; it was to create perfect order in the lives of individuals and in all society.

The later Greek philosophers and the Christians had sought perfection in the practice of *ethics*—the human pursuit of what is right and good for one and all. But modern science did not concern itself with ethics or with any other human values. Scientists were not interested in what they saw as vague and apparently religious ideas of what is right or wrong, good or bad, which could be argued forever and would only muddle the task of science. This they saw as the purely positivist task of describing natural mechanisms and passing their knowledge on to the engineers who would bring both nature and human society under control with perfectly designed and managed technology.

But neither personal nor social nor economic nor political problems were brought under such control. Science and its applications in linear cause and effect engineering made great advances in industrial production, in transportation and communications, in medical technology, in weaponry, and finally even in the exploration of space, but industrial nations were at one another's throats in the biggest wars ever fought and the environment was steadily razed to feed the machines. Enormous wealth, moreover, had been gained at the expense of vast numbers of the world's people, once self-sufficient, now poor, hungry, ill, uneducated, and without opportunity for anything better. The promised Golden Age of humanity seemed farther away than ever.

Meanwhile, scientists were extending ideas about the mechanisms of evolutionary change to the cosmos as a whole. Astronomers traced the universe back to a *Big Bang*—an original event explosion that was assumed to have created all the cosmos we know as its super-hot energy expanded. Stars and galaxies evolved, but moved ever farther apart, so that the universe as a whole was apparently spreading out and cooling off. Thus the great cosmic machine, the astronomers said, was running down—moving ever closer to its finish, when all order would be dissipated in the ultimate cold where no energy moved.

However far off this 'heat death' end might be, it was a depressing vision, and scientists offered no salvation from it, no comfort of values or ethics to give life meaning. People began seeing themselves as helplessly trapped inside a cosmic mechanism that would run down no matter what they did. Life simply had no meaning in this coldly scientific worldview, but who could oppose scientific knowledge? Modern philosophies such as existentialism and some schools of modern art reflected the scientists' view of a mechanical universe running down, humans caught in it like cogs in wheels, without meaning or hope.

Many scientists today still believe firmly in just such a mechanical worldview, but many others now see nature as alive and intelligent. Those who believe that life is self-creating in a dynamically alive universe, rather than winding down in a mechanical one, also believe that life can create its own meaning and purpose.

In Chapter 5 we spoke of the differences between mechanisms and organisms in connection with the autopoietic definition of life, and of entropy as the catabolic side of a metabolic cycle which builds up as it breaks down. But to really understand the present scientific debate on whether nature is or is not mechanical, we must go back once again to look at just what we mean by the concept and physical reality of mechanism, and at what role it has played in human history.

15

Less Than Perfect, More Than Machine

While the mechanical worldview and the explosion of technological progress it led to are historically Western innovations, their consequences in science, technology, economics, and politics have by now shaped the course of all humanity. Our invention and use of machines has become the guiding force of our species' evolution—we are now, for better or worse, technological creatures.

The word technology comes from the Greek *techne*, which originally meant any art, but has since come to mean the art of building mechanical systems, including our computer and telephone systems. Machines have given us powers far beyond those of our bodies, and we probably began inventing them to compensate for body parts that we lacked, such as long teeth, claws, and fur.

Our earliest machines—designed to extend the power of our hands and arms—were levers to move rocks, slings to throw stones, bows to fire arrows. To feed and clothe ourselves we formed flat and hollow stones for grinding and pounding food, spindles and simple looms for making cloth. All these are machines, as distinguished from tools, in that they have parts which move in relation to one another.

As our civilizations developed, we invented winding mechanisms and wheels, nuts and bolts, pulleys, and other ingenious devices for improving our machinery. We built mills and carriages and great machines of war to hurl missiles at enemies and climb their walls. But the real explosion of human technology came much later with inventions such as the printing press and the spinning jenny, which made useful things in larger quantities and less time than ever before; with inventions such as steamships and locomotives, which moved people about in larger numbers at greater speed than ever before; with inventions such as the radio and the telephone, which let more people communicate farther and faster than ever before.

Machines are made of parts that move to do something humans wish to do. In the first machines, the parts were moved by people themselves or by domesticated animals, so it is easy to see them as extensions of people. But as water power, steam power, fossil fuels, electricity, and finally atomic power were harnessed to machines, they seemed to take on a life of their own and we forgot that machines are still now, as they always were, a part of humanity invented by humans to extend human powers, rather than something independent of us.

Mechanisms come into being and function only through human design, manufacture, and use. They extend our power to build, to make things, to go places, to fight wars, to measure time and space, to perceive much more of our world and our whole cosmos in its tiniest and vastest reaches than can our senses alone.

Machines extend our power to amuse, teach, and talk to one another, to show ourselves to one another around our whole planet. They even extend our power to remember, to think, to predict and plan our future.

In all these ways and more, machines extend the powers of their designers and users. No machine would ever have existed without a designer and builder—not even the automatic machines that seem most independent of us. Science fiction writers may imagine worlds run by self-designed and self-reproducing machines, but machines will never exist without their creators and users somewhere in the background.

The idea that computer-run robots could come alive on their own is part of the misunderstanding even scientists have of mechanisms. Those who believe that life evolved by accident in a mechanical universe, on a nonliving planet, can also believe in accidents that will make robots come alive. But the fundamental distinctions between living organisms and machines show us why this will never be so.

Let us review those distinctions. Living organisms or systems remain functional only by continual change, whereas mechanisms remain functional only if they do *not* change, except as programmed. (Note that changes in natural systems can progress in only one direction, as they cannot undo their aging, while machines can run, in principle at least, both forward and backwards.) Living organisms are autopoietic and autonomous—that is, self-produced and self-ruled. Mechanisms, on the other hand, are *allopoietic* and *allonomous*— other-produced and other-ruled. The 'others' are humans, or human-programmed robots, which make other robots. A robot making itself by its own rules is a logical impossibility.

O   O   O

If we understand machines as extensions of ourselves and then think back to the mechanical worldview of Descartes, we can see that it was at least logical. Descartes understood natural mechanisms as God's creations—as engineered extensions of God's power, in the terms we used to describe mechanisms. It was later, when scientists decided to explain nature as self-evolving mechanics, without any creator, that a contradiction arose. Scientists were identifying non-created 'natural mechanisms'—which they believed to exist without purpose or design—with man-made mechanisms that exist *only* on purpose and by design.

Take, for instance, the human brain. When scientists took God out of their worldview, they also had to take out the idea of the human mind as a copy of God's. That left mind as a mechanism itself, or as the product of the brain mechanism. But just what kind of a mechanism could the brain-mind *be*? At first, scientists saw it as a kind of plumbing system—nerves were pipes and valves through which thoughts, feelings, and instincts flowed like water or got shut off and built up pressures that caused problems. When telephones were invented, the brain seemed more like a telephone exchange of messages along nerve-wires. No sooner were computers invented than the brain seemed to be a computer. More recently, some scientists have chosen to see it as a holographic camera and projector or as a parallel processor, which are among their newest inventions.

Now, in some ways, all these ideas were and are useful, for each of these man-made mechanisms—the plumbing system, the telephone exchange, the computer, the holographic camera and projector, the parallel processor—could be taken as a model of some aspect of the brain-mind in a way that would help us understand something about it. There is nothing wrong with using our mechanisms as metaphors for or models of nature—as long as we remember that they are only models and that they can only model certain measurable aspects of things found in nature.

The contradiction arises when scientists confuse the model or metaphor with the thing they are studying—when they believe, for example, that brains *are* complicated computers just a bit more sophisticated than present man-made ones, rather than seeing that computers are simply useful models of certain limited things brains *do*. Computers do these things in entirely different ways from brains, yet the model of the function can be valid in its limited way.

The confusion of models with reality comes from a failure to understand that scientists create *abstractions* the same way that artists do. If they did understand their models of nature as abstractions, they would no more confuse those models with reality than artists confuse their paintings or sculptures with the real subjects they portray.

But just what is an abstraction? To abstract means 'to lift out or away from.' An artist lifts out certain perceptions of something and makes them into a painting, while a scientist lifts out certain measurements of something and makes them into a scientific model. In both cases, the painting and the model are abstractions that stand for the whole thing. A mechanical bird can be considered an abstraction of a live bird into an assemblage of metal parts, just as Picasso's Guernica is an abstraction of human warfare into an assemblage of brushstrokes on canvas. Similarly, a scientific computer model of a biological or economic situation is an abstraction of certain measurable elements from the actual biological or economic situation. And, as an aside, many problems in our world today stem from what is not included in our economic models—such as the effects and costs of using up natural resources and polluting the environment.

Going back to the mechanical worldview of Descartes as an abstract world model, we see that he abstracted just those measurable features of nature that men were able to copy in mechanisms. Wind-up birds and church-tower puppets represented a few abstractable mechanical *aspects* of nature, but were then taken to stand for the natural creatures they represented. Still, Descartes recognized that theses mechanisms had to have a Creator—that they could not 'happen' or evolve on their own. Instead of seeing nature as autopoietic, that is, he saw it as God's allopoietic creation. Whether we think that a good description of nature or not, it was at least a logically complete system.

Now we can see that the danger of confusing scientific models with nature itself is that aspects of nature which we cannot measure, and therefore cannot abstract, may be the most essential aspects there are. Descartes' worldview, or world model, was logical because he understood that mechanical nature could not exist without an Engineer. But later scientists who dropped God from their explanations of nature failed to see that they were dropping the very essence of life from their world model. Much as they have tried to explain life in mechanical terms, their explanations have never been satisfying.

Scientists who do not mistake their models for nature readily admit they are only models. But they may still consider nature entirely mechanical by arguing that it is far more complex than, though in essence the same as, present mechanisms. More and more scientists, however, are dissatisfied with the mechanical worldview, recognizing that it is the self-creative aspect of nature that none of our mechanical models can account for. They are coming to realize that nature must be far more than mere mechanism, that it has a creative aspect no machinery can have.

O   O   O

Some ancient Greek philosophers, such as Pythagoras, had seen sacred geometry not as a human invention, but as the human mind's recognition of nature's underlying, designing intelligence. The mechanical worldview originated with a secular geometry that was pure mathematics, a human invention with no inherent consciousness. Such geometry, in and of itself is like mechanics in and of itself—it cannot spring to life any more than can a set of building blocks. Nor can secular geometry account entirely for such movements as those of the Sun, Moon, Earth, and other planets. Every calendar devised by humans has been plagued by the irregularities of nature. How much *more* difficult to explain by geometry the growth of an invisibly small egg into an entire human being!

The revival of ancient Greek sacred geometry today is proving valuable in explaining the fundamental physics of nature. Most physicists and mathematicians showing interest in it are those who understand consciousness as the source of creation and ever *inherent in* creation. This is a true revival of sacred geometry. Cosmic consciousness, in this scenario, assumes geometric forms to build a physical world in which they become an infinite variety of consciously self-assembling patterns—the improvisational dance described earlier, which repeats workable patterns in ever new configurations. Nature, we might say, is more an artist than an engineer, using the same recycled materials and the same schemes again and again, but endlessly creating something new from them and never machine-copying anything.

When we humans express ourselves through our technology, we usually copy some part of nature that we can abstract and translate from nature's evolutionary artistry into our relatively crude and lifeless engineering. But we must remember that our human ability to copy some aspects of nature in mechanical form does not in any way prove that nature itself is mechanical.

Let us go back to our earlier discussion of thermostats. The thermostat we install in a house so it will keep itself at the same temperature is a mechanical device designed to simulate what every warm-blooded creature does, and what we saw that our whole living planet does. But the 'thermostats' of our bodies and of the Earth are vastly more complex. Such natural thermostats cannot be removed from their living bodies, reduced to their parts, and rebuilt. They exist only in place as a feature of the whole body, and we can only search the intact body for evidence of their function, such as vasodilation and sweating.

We have copied the spinning and weaving of spiders, the termites' building of very tall structures, the trees' pumping of water against the pull of gravity, the tunneling of creatures into the Earth and their flying into the sky. We have copied the ability to see through darkness and detect things by sonar, to produce chemicals, solar power, and by now almost countless other natural wonders including the ability of our brains to solve problems. But though we can make mechanisms to copy things creatures do, we cannot even come close to building a working mechanical copy of the simplest single-celled creature as a whole. Our mechanics are limited in ways that nature's organics are not.

Does this mean that we must abandon mechanical models in science in order to understand nature? Not at all. We said earlier that the only way we can ever understand *anything* is by comparing things we don't yet understand with things we do—things that are familiar to us. And what can be more familiar to us than the things we ourselves have designed and created? If we had not invented the mechanical worldview along with our other mechanical inventions, we might not have made so much progress in understanding our world. But we must keep our minds open and recognize that nature is far more than mechanism, that we will hold up further scientific progress if we mistake our present models of nature for nature itself.

Descartes and the great physicist Newton built their worldview into a frame of space and time. Space and time were believed to have existed before the universe came into being, as a kind of stage on which atoms and the larger bodies they formed had been created and moved about lawfully. Each atom had a very definite location in space at any given time, in this model, and moved to new locations as time passed, according to fixed laws of motion. As the French astronomer-mathematician Pierre Simon de Laplace put it, an intellect that knew the positions of all the atoms in the universe at any given time could predict the entire future of the universe.

During the nineteenth century the discovery of electromagnetism and the new science of heat—*thermodynamics*, which later came to be called the science of complexity—shook this simplistic model. But perhaps the greatest blow to the mechanism analogy came when physicists were finally able to study the atom itself.

All atoms, remember, were supposed to be exactly alike, though they formed all the different things found in nature by being arranged in different patterns. Far too small to be seen, they were believed to be so hard they could never be broken or destroyed—they were thought to be not only invisible, that is, but also in*div*isible.

Even if non-material electromagnetism had to be added to material nature, and even if heat made things behave erratically, it was assumed that our ability to study the atom itself would surely confirm the mechanical worldview. Atoms, the smallest parts or building blocks of natural mechanism, must be moving lawfully in time and space. Scientists were at last ready to work out just how things were built from the bottom up.

But were they? We already mentioned the first shocking surprise they got—the realization that atoms were not all alike and were not tiny hard bits at all. Each atom seemed to be more like a tiny solar system, though its shape had to be guessed at from how it acted together with other atoms in forming molecules and their chemicals. Scientists often have to work this way to figure out the shape of things they cannot see with their own eyes. Think about our solar system—instead of being too small to see, as atoms are, it is much too large for anyone to see all at once. Its shape had to be figured out from the way the parts we *can* see act in relation to one another.

Anyway, just as the Sun is at the center of our solar system, something was at the center of the atom, with smaller things apparently whirling about it like planets. Physicists called the center of the atom the nucleus, and the things whirling around it electrons. Apparently, different kinds of atoms had different numbers of these electrons in orbit at various distances away from the nucleus.

The next surprise was that the atom's nucleus was itself made of parts, more tiny bits held together by forces so unbelievably strong that splitting the nucleus into its separate parts made an explosion. We all know what that discovery led to.

Every atom, no matter how tightly it is locked into its place—as, for example, in a crystal—turned out to be a tiny mass of jiggling, whirling parts. All the parts, around the nucleus and inside it, are nowadays called particles. But these particles soon proved not to be solid things either.

Deep in the very heart of matter, we now know, there is nothing solid at all. Particles are like tiny whirling winds in a storm of energy, or like waves dancing on a sea of energy. When physicists try to catch hold of them they rush off, leaving pretty curved trails. They disappear, divide, merge into one another, and reappear out of nothing—in fact they do anything *but* hold still to be studied. All the physicists can describe—or *try* to describe—is the pattern of their energetic whirlwind dance with one another—a dance that is, in fact, made of pure energy.

Such discoveries truly confused physicists, whose view of neatly ordered mechanical reality was shattered by them. Particles were neither solid nor reliable; they could pop in and out of existence with alarming speed and mystery. Einstein, furthermore, showed that time and space did not exist by themselves, as a stage for natural mechanisms, but were two aspects of the same concept and were really relationships created continuously as the universe created itself. Space did not even obey the laws of Euclid's geometry as had been believed, nor did time tick away in one great perfect clock rhythm. Instead, it seemed that cosmic spacetime curved like the dances of its tiniest particles, and one man could travel through this time-space without aging while his twin stayed home and became an old man.

The world seemed to dissolve at its very foundation. Yet it didn't dissolve into nothing, for the moving pattern of the energy dance is always there giving matter its form. It's just that separating the dancers from their dance—to study a particle as an object in itself— is quite impossible. As impossible as trying to take winds out of the air and waves out of the sea in order to study and understand a storm. If you try, you will find you have nothing at all in your hand—even though you know the storm is made of wind and waves.

Quantum physicists, such as Hal Puthoff, director of the Institute for Advanced Studies in Austin, Texas, research the zero-point energy field—a background of random, fluctuating energy that is everywhere, even in so-called empty space, even at absolute zero (whence its name) where no thermodynamics remain. It is now estimated that every point in spacetime, no matter where it is, contains—or is the source of—an infinite amount of such energy. If we remember Einstein's formula for converting energy to matter— $E = mc^2$—that means each point in our universe has far more than enough energy to create entire universes!

Puthoff found something extraordinary about atoms—that atoms themselves, always considered the most stable things in the universe, actually *lose* energy continually and must replace it from the zero-point energy source. This discovery means that our universe creates itself continually, not simply from a single Big Bang.

The atom itself seems more and more like the vortex or whirlpool we used earlier as a model of the simplest form an autopoietic entity could take—continually self-creating by taking in and spitting out matter/energy while holding its form. Rather like the giant protogalactic clouds that evolve into galaxies—the largest and smallest things dancing in concert to create our world and universe. Physicists such as Puthoff are now working to harness free zero-point energy as an alternative to fossil fuels.

The universe, after all is said and done, cannot be separated into parts as can a machine. Physicists have to be ever more inventive to learn about their strange new universe. They study particles, for example, in cyclotrons—the largest machines ever made, designed to allow scientists to study the tiniest things. But to see even traces of the particle dance, they must disturb it and try to work out what the real dance is like from traces of this disturbance. What matters, it turns out, is the pattern of the steps in the dance, for certain patterns of energy are what we call *matter*.

Dancers not dancing are no dance—and the dance, it turns out, is all there is!

Though we can never see the natural particle dance undisturbed, we can be sure it is there—forming and connecting the stars and their reflections in the sea, the Earth and all its creatures, ourselves and all the things we make and use. Everything is made of countless invisible dancers' movements in one single dance forming endlessly new patterns—a dance far too small to see and yet so large that it *is* the whole universe.

These discoveries, together with physicists' discoveries of larger dance-like patterns—patterns of wave mechanics in gases, liquids, and solids; of thermodynamics in heated matter; of electromagnetism—called for new ways of modeling nature, new kinds of mathematics that are less mechanical, more flexible, more like living nature.

All mathematics, up to the present, has been built on a foundation of mechanics devised at about the same time by Aristotle for logic and by Euclid for geometry. Yet, until the last century of the second millennium, the connection between logic and mathematics was not obvious even to mathematicians. Now that it is, mathematicians recognize logic—rules of orderly classification and combination of elements—as the true foundation of mathematics. And what this means is that mathematics can be changed as much as worldviews, because its logical rules can be changed.

By fiddling with Aristotle's logical mechanics, for example, mathematicians have created new and more dynamic systems of mathematics built on these changed foundations. Computers with their vast capacity for performing calculations have greatly increased possibilities for modeling self-organizing systems. Chaos theory, dynamics, complexity, fractals, sacred geometries sprout like mushrooms after a rain. In time, new kinds of logic, new ways of ordering human thought about a dynamically alive universe—an organic rather than a mechanical universe—will lead to a whole new kind of mathematics that will be useful in modeling such a universe. Science and mathematics are now working hand in hand on their exciting co-evolution.

Many of the new studies of self-organizing systems have been inspired by the work of Nobel Prize-winning chemist-physicist Ilya Prigogine, who revived the ancient concept of nature's creation of order from chaos, showing how self-maintaining systems even at a chemical level can re-create new order when they reach chaotic states. Prigogine's work extends the physics of equilibrium thermodynamics—which was invented to describe non-living systems—into non-equilibrium thermodynamics, which he used to model living systems. But let us keep in mind that his is still an attempt to describe living systems evolving in a non-conscious and non-intelligent universe.

We cannot repeat often enough that our scientific stories are changing more rapidly now than ever before. One particularly interesting thing to consider, is this: If the eastern philosophers were right in saying the world is illusion—that each of us creates our world from our beliefs—then what does it mean to measure a 'physical' universe with physical instruments? Are we measuring an illusion with parts of that very illusion?—creating ever smaller particles by believing in them? Our searches are leading us to fascinating puzzles and it is wise to keep very open minds in the process.

O   O   O

New theories and questions are part of our rapidly evolving scientific worldview. Yet we first encountered the organic worldview in the works of the earliest Greek philosophers, before the concept of natural self-creation was suppressed and God's perfect order became such an obsession that the whole Western worldview was changed to fit it.

Today's scientists are discovering things that many human cultures have understood—in their essence, if not in scientific detail—for millennia. Indigenous people who have not seen themselves as separate from the rest of nature consciously engage in its co-creation as one living system, using ritual, dance and myth as tools of their trade. Most important now is that western scientists—with their own ritual experimentation and theoretical stories—are coming back to this understanding of the universe as a conscious, alive and ever-creative dance of life. We ourselves are acting out its creative edge!

The ancient Greek myth told of Gaia's dance; the Indian myth told of Shiva and his wife Shakti, who forever dance the universe and our world into being. Of all creation myths, none tells of a world assembling itself mechanically as tiny parts come together to form the larger parts, which then come together as a whole world— none but that of our mechanical science, now passing into history. Rather, most creation myths begin with a whole—an undisturbed ocean generating individual waves, or a single being that divides into, or gives birth to, the different parts of the world. These parts may later rejoin as new wholes, or holons, within the great dance holarchy, in the repeating cycle described earlier of unity—>individuation—>conflict—>negotiation—>cooperation—> new level of unity. [See illustration in Chapter 2.]

We have seen that living systems are in many ways the antithesis of machinery; we have seen that images of dance fit many aspects of our new understanding of nature better than mechanical images do. To review, dance is a living, self-creative process as is nature in evolution. We may begin to create a dance spontaneously, as a natural expression of our energy that is not planned or designed in advance—as an improvisation. It may then evolve as new variations on the same basic steps create ever more intricate and meaningful patterns, just as in natural evolution.

Some of the patterns in a human dance may even be quite mechanical, as we express our mechanical ability through them, though we see and feel that the more mechanically perfect they are, the less lifelike they are. Classical ballet was an intentional effort to make dance as perfect as possible, and it developed at the same time our machine age developed. At the height of our infatuation with our machinery, while Chaplin was spoofing that, we developed ballets with long rows of dancers performing as nearly identical movements as possible—impressive, but never considered real art. The real art of dance seems to depend on human variation, on personal style, on imperfections, on surprise, to give it life and interest.

Classical ballet has become less popular than dances with freer patterns, and this may well be because our human search for perfection in the world and in ourselves no longer fascinates us as it did during the mechanical age. We seem to have satisfied our longing for perfection in building close-to-perfect machines. We want such machines to free us from our own boring, mechanical tasks, but we are rebelling against being treated as machines or machine parts ourselves—on the job, in schools, in government bureaucracies or wherever. We are tired of being told to be in perfect shape, in perfect control of our lives, because we begin to see now just how unnatural that kind of perfection is.

Nature is orderly without being perfect, as we have seen again and again. Nature's most useful patterns are never outdated but are kept for endless re-use, and the overall scheme of evolution is very stable and resilient. But mechanical perfection would be death to nature as it would be to us as part of nature. Nature is a live, self-creating process forever making order from chaos, forever free to do something new—to reorganize itself when necessary, even if only to stay the same; to create new forms when old ones no longer work. Perfection would be the end of evolution, the end of freedom, the end of creativity. We have learned that nature is far less than perfect for a very good reason—for the same reason that nature is far more than mechanism.

16

The Body of Humanity

The new scientific worldview we are forming is already showing great influence on our broader cultural worldview. Just as mechanical images inspired the development of industrial and social technology, organic images of self-creating networks are beginning to inspire us to reorganize all human society as a more harmonious and humane venture.

Gaian evolution itself is pushing us in this direction. The evolution of a worldwide body of humanity is very much a step—in fact, the newest step—in Gaian evolution. Like the rest of evolution, it was not planned, but is free to occur and consistent with the overall pattern of the dance.

Much as we humans have been creating this step through our technology, we have not been creating it intentionally any more than we intended to destroy our environmental life support systems as we created our industrial lifestyle, or any more than we set out to create a means of committing species suicide when we invented nuclear weapons. We have just begun to understand that these are the real or threatened consequences of recent human activity and that they put our very survival in question.

In just the past few hundred years of our half-million years as tool-making humans, we have used our big brains and clever hands to produce a technology that changed the whole planet and united us into a new kind of being. Without strife, we have built an efficient worldwide system of mail, telephone and electronic communications, a worldwide air, sea and land transport system, a global money exchange, a United Nations with many cooperative agencies, and a vast system of non-governmental organizations.

Our multinationals are global, we have a World Trade Organization and a World Parliament of Religions; we are continually working on international agreements of all sorts. Yet we have hardly even been aware that we were evolving into a single body of humanity. It happened as naturally as the evolution of our physical bodies.

Most of our understanding of ourselves, of our evolution, and of our social history has, after all, been gained only very recently. Before this century, we couldn't even know what was happening in the rest of the world *while* it was happening, much less trace its roots into the dim past. Quite suddenly we live in an age of telescopes that show us the most remote parts of our universe and its most ancient history, an age of microscopes that let us look deep into the tiniest parts of our own bodies and the rest of nature. Only in this age have we begun digging up the fossils of our early ancestors and the remains of the first human civilizations, making them into books and films that tell an ever more connected and meaningful story.

Only now can we see our whole planet from space and begin to understand it as a great living being. Only now do we see that, from Gaia's perspective, life evolves as a whole—rock transforming itself into what we perceive as a great variety of separate species, as well as into what we see as the various environments of land, sea, and air. But, as we have seen, environments are not lifeless geological habitats in which living species evolve; they are themselves collections of living species and their products. Sea, soil, atmosphere, and even hard rock are all products of Earth's geobiological metabolism as a live planet.

If all creatures and environments co-evolve by changing themselves and one another, then to understand any particular species we must try to understand how its evolution is related to the evolution of its environment. As we said, rabbits cannot evolve without their habitats and vice versa—all we have is rhabitats. In particular, we can only understand ourselves as humans by trying to understand our co-evolution with the rest of nature.

Let's go through the story of our human evolution just once more, recognizing that its pre-historical phase is still very murky and that alternative stories are more plausible to some of us. Let's imagine seeing this evolution from a distance and sped up as a short film.

First we see small groups of humans evolving in dense forests in the warmer areas of the Earth. The climate changes and the forests shrink—we see them walking upright on the ground, groups of them wandering in search of food, some finding permanent shelters in caves and other protected places. Using the resources of their environment, they begin to make things that are of use to them, things that compensate for their lack of fur, sharp claws, and long teeth; things that help them hunt other large animals for food, bone tools, and clothing; things that help them carry, store, and prepare food. They learn to control fire for warmth and cooking, and to carry live embers from place to place. When their families or tribes grow too large to live together easily, some members bud off to form new tribes.

The human creatures thrive, multiplying and spreading out as they follow food and water supplies. Great ice ages push them back to warmer climes, but each time the ice thaws, they are lured toward the lush new growth springing up in the wake of the ice. Eventually their food supplies draw them to all the continents. Some remain tribal hunter-gatherers or nomads while others begin the settling process we call civilization.

In the best climates, groups of them settle to make houses, villages and gardens, to keep animals and grow crops, to store food for dry or cold seasons. Some begin making boats to explore along rivers and venture out to islands in larger bodies of water. Villages grow into towns, and towns into larger agricultural societies that transform considerable areas into manmade ecosystems. Barter among settlements and wandering tribes develops an economy of exchange. For thousands of years they bud off new nomadic tribes and settled colonies as numbers slowly expand and they spread over the habitable areas of the world, developing their arts of plant selection, animal husbandry, pottery, painting, and metalwork.

Then we see the larger agricultural economies overrun by tribes of wandering nomads and hunters from harsher climates, armed with weapons, taking them over, establishing dominance systems of males over females, rulers over those ruled. They build kingdoms and unite them into empires through warfare. More and more land is taken for human use. The old self-creating, self-balancing ecosystems are destroyed as natural plants are cut or burned, their animals driven off, both replaced by human-bred monoculture crops and livestock, as well as by walled cities of stone and brick.

Within and between empires, wars are fought and goods traded, building networks of land and sea paths that connect human societies with one another. Along these paths, news, ideas, and stories flow together with people and their products, animals, seeds, microbes. Sometimes unwittingly, people change whole ecosystems as seeds or animals they import take over and drive out the native species. Cities, in which natural land is replaced by man-made buildings and streets, grow up as centers of ideas, inventions, new ways of life. Their crowded conditions also breed disease—plagues sometimes wipe out whole populations. Natural disasters parch their croplands, flood or bury settlements, reminding them of nature's power.

The borders around kingdoms and empires change; continents are mapped into countries; human populations grow and divide into ever more languages and cultures. Ecosystems have shaped human civilization by drawing it to favorable climates, into fertile river valleys, along coasts, and wherever humans find the easiest overland transportation routes. In turn, humans transform the environment ever more to their use, especially by cutting forests for the use of wood and to clear land for crops; by breeding and herding hoofed animals that eat vegetation down to its very roots. Humanity proves to be a desert-making species—to the extent that the deserts it creates are the only sign of human existence visible from the Moon to this day!

Cities crowd more and more people together in artificial environments; raw materials are transported to the city centers from more and more distant places, while the products manufactured from these materials flow outward again toward markets. Crops and animals native to one part of the world are planted and raised in others.

Human technology evolves from horses and sailing canoes to steamships, jet planes, and spacecraft, from weaving looms to computer industries, from town criers to television. A twinkling cobweb of electric lights lights a world once dark by night except for forest fires. A world once silent by night except for the lone cry of a bird or mammal is filled with the sounds of machines and music. Mines and quarries have been dug deep into the Earth and scratched out of its surface, their stone, metal ores, and fossil fuels transformed into human products.

Rivers have been dammed up and diverted into unnatural paths, flooding ecosystems behind them, making deserts in front of them, for the sake of the insatiable human demand for electrical power. Whole forests have been cut for lumber and fuel or burned to clear land for grazing and agriculture. More and more natural land is plowed under by farmers and paved over by builders of cities. Deserts grow larger while more and more species of animals and plants are killed off as humans exploit nature for their own purposes.

Man-made fertilizers, pesticides, heavy metals, and other waste materials of human production pollute the atmospheres, the waterways, the soil, and the oceans. Yet food production and other technologies have suddenly exploded humanity itself into vast numbers with rapacious appetites for food and energy, destroying and outstripping the resources Earth can provide.

Wars are fought on an ever-larger scale, ever farther from home, and involving greater numbers of people spending longer times in strange countries. A holocaust shocks the world with its unthinkable but true atrocity of man against man. Yet war brings people together in positive ways—soldiers stay to make friends with and marry their former enemies, raising children together; others leave their war-torn countries to adopt new ones in seeking a better living.

Wars drive technology and industrialization to new heights, especially through the development of an enormous fossil fuel economy that spawns vehicles on land, airplanes in the sky and ever more ships crossing seas. More resources are dug and stripped from the Earth than ever before. Nuclear energy is a product of war—two atomic bombs are blown up in warfare, deliberately destroying a part of man's own civilization. Others are blown up as tests, destroying and polluting ecosystems, raining fallout from the atmosphere worldwide. The peaceful use of nuclear energy proves dangerous as well, with accidents creating radiation sickness and damaging foodstuffs. But the war that gave birth to the use of nuclear energy also gave birth to the widespread use of computers and the ability to create an Information Age to succeed the rapacious Industrial Age.

Despite an intense Cold War after the two hot ones, in a world divided into Capitalist and Communist camps arguing which of their systems is best for the world, more and more people swell transportation systems as they are sent to work and live in one another's countries, or as they choose to go there on holiday. They learn one another's ways, sharing more and more ideas. Cultures are mixed within political borders; cultures are shared through networks of local and foreign communication; ever-larger numbers of people become literate and learn what is happening in their world. Even people who never set foot in another country can eat and use the whole world's products and know the whole world's ways of life in full sound and color.

People prove that they are capable of mingling and sharing, yet governments maintain hostilities. Artists and scientists try to bridge the gaps between hostile peoples, to share their works and knowledge, their fears and hopes for humanity. The threat of nuclear holocaust has driven even politicians to seek new ways of working out differences. The old separations of distance, language, and culture are bridged as the human technologies of transport and communications bind humanity inevitably into a single worldwide body. But that body is plagued by the vast numbers of people who have been dispossessed in its building, who go to bed hungry and ill, who die as children.

Human technology makes the leap into space—for the first time ever, we see our exquisitely lovely planet from afar, as a living whole. Humanity suddenly awakens to the recognition of the vast damage it has done to its environment and thus to itself. It is beginning to understand the threat of exhaustion or irreversible pollution of natural waters, fossil fuels, and other supplies; to recognize its power to destroy the whole human world and force the planet into new paths of evolution; to feel the effects of its greenhouse gases in an atmosphere that is growing uncomfortably warm and could kick the planet into an ice age or worse, a hot age. Just before the second millennium ends and the third begins, scientists recognize human effects on the planet as its Sixth Great Extinction—an extinction progressing more rapidly than any before it, even that caused by sudden meteor impact sixty million years ago. It is the first extinction caused by a single species. We see the enemy now, as Pogo told us, and it is us.

O   O   O

On the other hand, an indigenous wise man, a Hopi Elder, tells us that "We are the Ones we have been waiting for." *We*—not some imagined rescuing Savior—are the only ones who can turn disaster into opportunity; *we* are the ones who can understand our interconnectedness in the great web of life and our power to honor it, treat it as sacred, cease damaging it, restore it. Will we understand that in time?

The history we have just watched is an impressive scenario, its saga ending on a frightening note. One species—one new upstart species—has appropriated the entire planet to itself, turning rich and varied ecosystems into fragile monocultures, vast deserts, and choking pollution. Are we a kind of planetary cancer, looking heedlessly to our own expansion at the expense of our own support system? Why is the only species with so much capacity for hindsight and foresight so destructive?

Let's be brutally honest with ourselves, for if we are the ones to change things, then we must look squarely at ourselves. The most obvious feature of human social, political, and economic systems continues to be empire building through dominance: humans dominate other species; the female half of our own species is still largely under the control of and exploited by the male half; most of the Earth's countries are still dominated and exploited by the few most powerful ones or by the banks and multinationals they have created; individual countries maintain their own dominance systems of class, caste, and discrimination, the few hiring the many to work for them and bring them the financial advantage that drives our economies; vast numbers of people have been dispossessed by this domination and driven to abysmal poverty and ill health; wars continue to erupt as dominance over land, resources, and beliefs are contested; the dominant culture is eradicating natural and cultural diversity.

Why this pattern of dominance, of competitive exploitation? Are we unique, or is this a normal stage in our species evolution?

We have already compared the evolution of the body of humanity with similar events earlier in our planet's evolution, suggesting that the development of communications and transportation, and the shift from competitive exploitation to a cooperative division of labor, are comparable to earlier processes—ancient bacteria evolving into protists, protists evolving into multicelled creatures, ants evolving into ant colonies, and so on. All these show us a pattern repeating now, as modern countries evolve into a worldwide body of humanity—a new *global* multi-creatured cell.

Fractal Biology

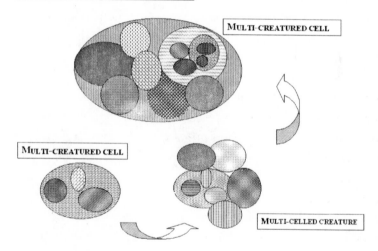

In this comparison, the body of humanity is not fully evolved, because exploitation and dominance still grow side by side with cooperation, and one of their most dangerous effects is the ability to destroy diversity in favor of monoculture—in energy production, in agriculture, in cultural fashions. The globalization of an adolescent American CocaCola and pop music monoculture that destroys all other cultures through the seduction of their own adolescents is no more viable than the genetic engineering of identical plants and animals. The vital importance of diversity to effective cooperation in nature has yet to be clearly recognized by the dominant human culture.

Surely there were far more failures than successes as ancient bacteria evolved into protists—countless instances in which unceasing exploitation and hostilities among bacteria multiplying within a single cell wall led to the destruction of the whole enterprise. Perhaps, in a parallel fractal way, globalization struggles to happen on countless planets in our universe that have evolved civilizations, but we humans cannot afford to be one of the failures, as we have only one chance—the common cell wall that binds us together is the boundary of our planet itself.

If we understand the evolutionary pressure on us now to complete the organization of this new body, we can work at the task consciously and rapidly. To see more clearly what needs to be done to complete our organization into a single healthy organism, let us look again at the successful evolutionary precedents of eukaryote cells and multicelled creatures.

The organization of the bodies of multicelled creatures—including us—is much like the organization of eukaryote cells, except that organs take the place of organelles, a brain evolves instead of a nucleus, blood and lymph vessels instead of cytoplasmic transport channels for supplies and wastes, and so on. Since most of us are more familiar with the workings of our own bodies than with the workings of single cells, it may be easier to see the relationship between our individual bodies and the whole body of humanity than to keep talking of cells.

Let us, then, play out this metaphor, or analogy—this comparison of our familiar bodies with the still unfamiliar great body into which we are uniting—by regarding countries and multinational corporations as organs, by seeing shipping routes that carry supplies and products as blood and lymph systems; communications networks that spread information and ideas as a nervous system; and attempts at building some kind of world government as first steps in the evolution of a brain that can coordinate all the body's activities. And let us acknowledge that the Gaian experience accumulated in the evolution of our bodies, as well-functioning and representative living systems, is worthy of respect.

Consider economics and politics—the ways in which we manage our products and ourselves. How do we organize these basic functions of collective humanity and how does this compare with the organization of basic functions in our physical bodies?

Economy—the way we organize the making and shipping, the selling and buying of our human products and services—meant 'rules of housekeeping' back when the word was coined and everything people ate and used was grown or made within households. Now our human household includes all of Earth and we might call economics our 'operating principles' and ecology our 'organizational design.' Our economy is a worldwide system of manufacture and trade that works by both national and international rules. Yet this system did not evolve to serve a worldwide household at all—it was not intended to become a single system. It grew out of rather lawless competition among individual nations, though it was eventually forced by its own evolution to make international rules for managing it. Unfortunately, these rules still serve the interests of those who already have economic advantage better than those who do not.

The industrial countries that set up the international economy, with its World Trade Organization management, simply have more money and power to make political and economic decisions than do the poorer countries that supply their raw materials and cheap labor. If we continue the analogy with our own bodies, we can easily see why this is an unhealthy situation. The parts of our bodies—its 'nations'—work together as organs and organ systems, such as bone, blood, muscle, and digestive organ systems. If all these organs and systems did not work harmoniously within themselves and with one another, our bodies couldn't function.

Imagine, for example, what might happen to us if our bodies' economics worked like the economics of human society. Raw material blood cells are produced inside bones all over the body, just as raw materials are produced in supplier countries all over our world. The raw material blood cells are then transported to the 'northern industrial' lungs, where the blood is purified and oxygen and nutrients are added, making it a useful product.

So far, so good. But imagine the announcement of the heart distribution center, "Today's body price for blood is such-and-such. Who will buy?" Some of the bones in which the raw material blood cells are produced can't afford the oxygen-rich blood they need to stay healthy. But rather than lower their prices, the industrial organs destroy the surplus blood that no one can afford to buy, or put it in storage, hoping to sell it later. Bone cells begin to die of starvation. The starving bones would soon affect the whole body, making it unhealthy, crippling or even killing it.

It is clear that a few organs cannot exploit the rest of the body to their advantage. Nor do we find families that starve three children to overfeed the fourth. When we think of our bodies or our families, we have no trouble understanding why all their parts must be healthy. Yet, we do not manage our national or global economies by this same wisdom.

Even though our products, including our food, originate all over the world, we do not share fairly the means of their production or their distribution. The UN tells us that our food supplies are presently enough for all humans to eat well, but industrial countries own or control the bulk of food supplies, and they can set prices for the world market. Rather than let prices go down by flooding the market with food, they hoard or destroy surplus food and pay farmers in their own countries to stop producing, while huge numbers of humans go hungry.

Countries that grow food crops for export to industrial nations often do not grow enough food for their own populations, many have starving people. Bangladesh and the Sahel countries of North Africa, for example, have suffered starvation during years in which their food exports were at their greatest height. So much of their productive agricultural land is devoted to export crops that the very people hired to grow these crops have insufficient land for their own crop needs.

It is for reasons such as these that our news media often report starvation side by side with 'crises' of overproduction! The solution, except in times of emergency, is not to give away surpluses to the hungry, but to redistribute the arable land so that they can feed themselves.

In our bodies, troubles of this kind do not arise, for our bodies evolved cooperative economic systems from the start. Illness or injury can of course stem from outside sources or from internal breakdowns, but our brains quickly detect such problems and see to it that any part in trouble gets immediate help from other parts. In its natural wisdom, our body recognizes that any unhealthy part threatens the health of the whole. It is no doubt fortunate that our everyday consciousness is not in control of such matters, for we have proved, at least so far, much less wise than the 'automatic,' unthinking parts of our brain that coordinates body affairs. We shall return to this observation in the next chapter.

It is obvious that a living body can be healthy only if its systems function cooperatively. As long as human economics remains more competitive than cooperative, we hold up progress toward the evolution of the body of humanity.

The problems in our world economy have become even worse because starving populations—strange as it seems—grow faster than well-fed populations. It is as though the bone cells of our bodies, seeing their kind dying off from starvation, produce ever more of themselves in a frantic attempt to preserve life. Poor people see their children dying and make more. It is natural for them to love their children and to feel that the more they have, the better off they will be. More children mean more workers to bring in family income and care for parents in their old age. Over-population began when the colonial process broke up the communities of self-sufficient and self-regulating populations, as will be further discussed in Chapter 20.

The well-fed people of richer countries do not have such worries, and they have the opportunity to do many interesting things besides raising children. There is no overpopulation problem among rich people—on the contrary, some rich countries have an under-population problem. It has become quite clear that if everyone in the world had plenty of food and opportunity, we would not have developed a population problem. Yet our efforts to solve this problem are all based on curtailing populations by law and contraception rather than by recreating self-sufficient communities, raising living standards and increasing opportunity.

People who are not hungry are also less angry. Much warfare in our modern world is a result of conflict between rich and poor—

poor people trying to get back land and resources taken from them in colonial times by industrial countries, industrial countries trying to keep or get back their control over these sources of raw materials.

Which brings us to politics.

O   O   O

After the Second World War, the most powerful nations divided themselves into two camps embracing the competing political-economic systems of capitalism and communism, as we all know. One side said people should look out for their personal interests and the whole society would flourish naturally through their competition, as in Darwin's theory of evolution. The other side said people should cooperate by sacrificing their personal interests to work for the good of the whole society.

The differences between capitalism and communism actually proved to be a good deal less sharp in practice than in theory. Both systems had essentially the same industrial structure: bosses and workers filled communist as well as capitalist factories and lived basically similar lives on both sides. Both sides recognized to some extent that their own people could no more afford to ignore collective society's interests than collective society could afford to ignore individual interests. Unfortunately, they did not extend this recognition and practice to their international politics.

What their international politics were really all about was the struggle for power—especially the power to control the cheap raw materials of the less developed world to feed their industrial processes, as Alvin Toffler well described, and as David Korten pointed out more recently. Each side concentrated power and disempowered its people, wasting much of its human potential. And each side claimed its system would be best for the whole world to adopt and did whatever it could to push or persuade developing countries into their camp.

The competition of the big powers—the United States and the Soviet Union—for the allegiance of the rest of the world fanned political conflicts and outbursts of warfare that periodically threatened all humanity with their escalation into global nuclear war. Both major powers made enough nuclear weapons to destroy their opponents as well as themselves and seriously damage the rest of the world, including other species. By the time the Cold War ended, their nuclear and germ warfare technologies had spread to China and many smaller nations, creating an ongoing threat. The international trade in arms has become an enormously profitable enterprise. In fact, worldwide revenues from the sale of arms and drugs exceed the entire budgets of many nations.

Everyone knows by now that there is no way to fight a nuclear war without bringing on catastrophe for both sides as well as for those not involved in the conflict. If the body of humanity continues evolving, rather than destroying itself, we will see ever more nuclear disarmament agreements. But disarmament will not be enough to bring peace and equity, for there is another danger perhaps even greater.

The human mania for making monocultures is apparent in our social behavior as well as in our agriculture, because we simply have not recognized the vital importance of variety or diversity in any natural system. No such system or body could function if some of its species or organs had the power to make the other organs over in their own image. Imagine just a single such circumstance—imagine your heart trying to persuade or bully your liver into being just like it. Its success would clearly be a disaster for the body as a whole. Do we *really* want the Malaysians or the Inuits or whoever is not like us to become just like us? Nature makes it abundantly clear that the secret of success is mutually cooperative variety.

The fact that humanity divided itself along lines that promote individual versus collective interest, is not so surprising when we look at nature. This conflict between individual and collective interests preceded the organization of protists from individual bacteria and the organization of colonies and multicelled creatures from protists. In fact, the whole Gaian system must constantly work out this conflict.

It is clear that every natural creature from paramecium to plum tree to puma looks out for its own interests by feeding and protecting itself as best it can, just as Darwin said. What Darwin failed to see was that nature is not made only of competing creatures in backdrop environments, but rather of those living holons we know as wholes in themselves, that are also parts of larger holons—all nested into holarchies. Now, if every holon at every level in such an arrangement looks out for itself, we have a situation in which selfishness continually transforms itself into cooperation!

How this is possible can be seen easily by considering our bodies once again, and in the next chapter we will show examples elsewhere in nature. Each of our cells is a living system, or holon, in its own right. Yet, as a holon, it is also a part of larger holons—organ, organ system, and whole body—together forming a physiological holarchy. Clearly every cell manages to look out for its own interests—to care for itself and to reproduce itself. But the organ it is part of *also* has self-interest, as does the body in which the organ resides. So we have a situation in which there is self-interest at *every* level of holarchy.

Two things can happen in this situation: one is that some level gains the power to destroy other levels to meet its self-interest, in which case the system will break down, as we saw in the example of blood distribution; the other is that self-interest at every level leads to negotiations that bring about cooperation and well-being in the whole system. This should remind you of our evolutionary pattern: unity—>individuation—> negotiation—> cooperation—>unity.

Darwin saw evolution as driven by competition among individuals. Later evolutionists noted cooperation and altruism within species, suggesting that evolution must be driven by competition among species for ecological niches in which to flourish. Richard Dawkins then proposed that both these theories were incorrect, as evolution was really driven by selfish genes that struggled to maximize their expression in the overall gene pool. What none of them had the vision to see was that they were all right, but only *together*. Self interest at all levels—species, individual and gene—motivates nature's creativity and health.

Even a couple learns that couplehood has interests in its own integrity apart from the interests of either partner. That is, a couple is a two-level holarchy, with levels of individual and couple. As the ancient Greek playwright Aristophanes put couplehood dilemmas: "You can't live with 'em and you can't live without 'em." If the couple has children, they become a family with a new level of holarchy, which is embedded in a community, and so on. It is very important to recognize that self-interest is not a bad thing, *except* when it is not contained and modified by negotiations with other levels of its holarchy. This clearly suggests that a world economy can work well only if it recognizes the need for strong local economies within it, rather than destroying them.

Nature works out dynamic balance between self-interest and interest beyond self, as we can easily see in our bodies. It is no doubt for good reason that every cell in our bodies contains the gene plans or resources for the whole body, since most of our cells must stay in place and thus cannot take what they need from a common gene pool when they need it, as do the streamlined free-flitting bacteria. The genetic directions, or resource libraries, in all our cell nuclei may even be organized as a holarchy of holons representing interests at all levels from the whole body to those of each individual cell. This is speculative, yet we do know that our whole bodies are clones of one cell and that each cell switches on the genes that concern its particular organization and work. What is speculative is whether each cell is in some sense directly informed by its nucleus about the rest of the hilarity's needs. If that information is *not* in each cell and continually updated, then it must be available through non-physical communications among all cells. Otherwise our bodies could not function.

We know that there is communication among cells and that each cell's organization and work are related to that of its organ, its organ system, and the whole body. This entire system unfolded during our embryonic development in such a way that each level of the physiological holarchy from cell to body looks out for its interests, and thus they are pushed or pulled into cooperation.

If every cell in an organ worked for its self-interest, but the organ as a holon did not, the cells might well kill one another off in competition. Surely they would be disorganized to the point where there was no functional organ. In the same sense, a society in which people looked out only for their individual interests, because they were not asked to do anything in the interest of their collective society, would not be a functioning society. This is why capitalist societies do have governments to manage the public interest, to create public works and institutions, to limit free enterprise and tax some of its profits to meet society's needs.

Consider now the opposite situation, where the organ or the society is so powerful a holon that it can demand the complete self-sacrifice of its cells or people in serving its interests. The cells or people thus enslaved would no longer be individuals in their own right. Science fiction writers have tried to imagine humans becoming robot parts of a mechanical society, but people complain bitterly about being cogs in a wheel, and they stop functioning well. This is why communist countries have either failed, as in the Soviet Union, or discovered they must give people some opportunity to work for their individual interests if their societies are to work as a whole, as in China.

Capitalists were right that people must work in their own interests, and communists were right that society must work in its collective interest, but both were are wrong in claiming that one or the other will do by itself. The present worldwide shift toward free-market capitalism will work in the long run only if it incorporates the best aspects of socialism—the concern for the whole as well as the parts, including concern for the welfare of the entire body of humanity and its planet.

Nature never requires any individual to choose between its own interests and that of its larger body, society, or ecosystem as humans have been doing in forcing such choices, as we did between capitalism and communism. With the breakup of the communist world, it becomes ever more important to heed Toffler's advice that we stop looking at every idea in terms of whether it comes from the left or the right and see instead whether it takes us forward or backward. And the best way to see that is to look at living systems and how they function when healthy.

O   O   O

The body of humanity has not yet evolved the truly impartial and cooperative world government it needs to coordinate its interests as a whole. Looking back at evolution again, we recognize that there must have been a number of steps in the transition from monera to protists as competition among individuals gave way to their cooperation as members of a new whole. We know that one of the most important steps was the formation of the protist nucleus from the DNA of the various monera living within the same cell walls—the nucleus that could coordinate the information needed to carry out the activities of the whole. The same step was accomplished when nervous systems formed in multicelled animals that had evolved from protist colonies in which different member cells did different jobs.

Something of this ilk is clearly happening as the body of humanity struggles to form its new identity. Since the close of World War I, people have recognized the need for some kind of organization to coordinate and balance national and international interests. First they tried the League of Nations, then the United Nations. Although the UN has accomplished much in the way of programs and services, the competitive interests of member nations still dominate on important issues, limiting the powers and often preventing the smooth functioning of the UN. Some powerful organizations spun off from the UN—such as the World Bank, the International Monetary Fund and the World Trade Organization—clearly serve the interests of powerful nations and multinationals over the poorer nations. In this situation the wealth of the world is ever more unfairly distributed and the gap between rich and poor grows dangerously wide.

The rise of official UN NGOs—UN-affiliated non-government organizations, many of them grass roots based—is an interesting development. It remains to be seen whether they will be incorporated into the present UN structure as it is reformed, or whether they will become a kind of parallel UN and history works out which will become the main organization. A world government, if it follows evolution's lessons, will not be autocratic or authoritarian but will become a world government in service to the needs and welfare of the body of humanity, as are our brains and nervous systems in our individual bodies. If our human civilization is to survive, we have no choice but to solve this problem before long, completing our evolution into a worldwide body of humanity with a functional coordination system.

We must also ask: How can the body of humanity function if half its cells suppress the full expression of the other half? It is a blight on humanity that neither the UN nor any single country in the entire world, not any multinational corporation, has yet paid more than lip service to training and selecting women for half of its governing and professional positions. Nowhere is it recognized that such equality may be fundamentally necessary to the health of any society, that a system of sexual inequality inevitably breeds conflict while losing valuable resources and justifying every other form of inequality, oppression, bigotry, and antagonism.

Our biggest job is to change our whole way of thinking to a larger perspective, to recognize ourselves as a body of humanity embedded in, and with much to learn from, our living parent planet, which is all we have to sustain us. How can we as a species live in harmony within it? How can we as people live in harmony within our own species?

The sooner we recognize ourselves as being in transition from exploitative and divisive practices on all fronts to a united and harmonious living system, and the sooner we recognize that there are natural models to guide us, the sooner we will complete our healthy evolution by our own choice and efforts.

17

A Matter of Maturation

We have seen human worldviews change dramatically from the early view of nature as the Great Mother to a view of nature as the mechanical creation of a Father God, then to the portrayal of nature as mechanism evolving by accident, without purpose or design, there to be used for human purposes.

A psychologist might see this sequence of worldviews that is the heritage of industrial and post-industrial humanity as having something in common with those of an individual member of our present society passing through stages of personal development. Technological culture, on which this book is focused, has clearly become—for better or for worse—the dominant human culture that will make or break us as a species. What matters, then, is to recognize that this culture is still immature from a developmental perspective. We may then hope it will learn not just from its own experience but also from that of the few remaining non-technological cultures it is wiping out in its drive to so-called progress, though they may be far more mature in their relation to our parent planet. In Chapter 19 we will look more deeply at who they are and how we might cooperate with them fruitfully at this critical transition time for humanity.

Of our species infancy we know very little; our earliest artifacts, as we said in earlier chapters, indicate a recognition of nature as mother and of our closeness to and respect for her. In our long, relatively peaceful childhood we learned nurture from her, developing agriculture and art. Father figures existed primarily in absentia from the early agricultural civilizations—as the gods of nomads and hunters who eventually came to disrupt them with violent conquest and lasting domination.

Under the influence of paternal gods, we formed our *ego* the Greek word for 'I'—coming to see ourselves as separate from nature, growing out of the close union with nature-as-mother into seeing nature as separate from us, as the creation of an authoritarian Father God, in whose laws and demand for obedience we found some security.

In this analogy, the European Middle Ages seem to be our prepubescent phase—a stable God-fearing Christian society that lasted over a thousand years under His authority—until man began expressing his ego more boldly. In time he challenged religious stories, or revelations about the world with scientific observations, making discoveries and developing technology in ways that permitted him to transform nature ever more to his own purposes. Was it not as if humanity through its Renaissance and Enlightenment reached the stage of competence and confidence we associate with adolescence? As scientific knowledge and technological industry swelled the adolescent ego, the father's authority was rejected and nature was seen as no more than a knowable and predictable environment for men to control and exploit as they wished.

Is it not to be expected that a smart and clever adolescent will reject or at least question the unchallenged parental authority in which the child believed? Is it not common to gain in adolescence, between bouts of insecurity, the conviction of knowing everything and being in full control? Why shouldn't whole human societies go through the life stages of childhood and adolescence as each individual human does? Is not our whole species, quite like every child born to it, still young and free to learn from experience?

In mythology—mythology long having served as cultural psychology—the heroic cycle often represents the life cycle. The youthful hero leaves home, encounters challenges in the course of his adventures, then finally returns as a mature, wise man. Such myths often portray the hero as a brash youth with the *hubris*—the gall, as we would say—to believe himself as invulnerable and powerful as the gods themselves. For this he is invariably punished by a fall, which may be permanent if he does not learn the lesson of false pride.

In real life, the adolescent who strikes out with a false sense of maturity, believing he or she knows it all, can be expected to get into some kind of trouble before maturing into an adult. And the adolescence of civilized humanity is running true to form. Our view of nature as a mechanism to be exploited fostered great progress in technology, but we made this progress recklessly in our belief that all nature was ours to do with as we pleased. Now we find ourselves punished by the enormous problems we have created along with our modern technology.

Like any adolescent who is suddenly aware of having created a very real life crisis, our species faces a choice—the choice between pursuing our dangerous course to disaster or stopping and trying to find mature solutions to our crises. This choice point is the brink of maturity—the point at which we must decide whether to continue our suicidal course or turn from it to responsible maturity. Are we going to continue our disastrously competitive economics, our ravaging conversion of our natural supply base into things, our pollution of basic soils, waters and atmosphere in the process? Or will we change the way we see life—our worldview, our self-image, our goals, and our behavior—in accord with our new knowledge of living nature in evolution?

Will we come to hold nature sacred once again, as wise indigenous elders urge, so that seven or more generations to come will benefit from our decisions? We are at the point where we can see our own historical evolution and decide whether to hold up its natural advance into maturity—prolonging our adolescence dangerously—or whether to speed its course by making haste in the face of crisis to complete the mature cooperative body of humanity by conscious choice.

Growing up is not easy, as we all know. Youth must fall on its face in its ambition, must learn by experience; the hero must be wounded in battle and be knocked down for his hubris, his pretension to godhood. Maturity comes only when youth gains perspective on itself and is at last willing to admit there is still something to learn.

As we have not yet gained this perspective, many of us believe that today's human problems will never be solved, that they have simply gotten too big for solutions of any kind or that, even if we solved them temporarily, human nature cannot itself change and therefore we would just get into the same mess again. This pessimistic view of ourselves as a species reflects the way we feel as individuals whenever we are depressed and our problems seem insoluble.

Hopeless pessimism often comes from lack of perspective. If we look at things narrowly, from within a difficult situation, they may well seem hopeless, but if we manage to step out of our dark hole, so to speak, to gain some perspective on ourselves within it, we may begin to see a way out.

The purpose of this book is to put human life into just this kind of perspective—to see ourselves within the whole evolving world, even within the whole evolving cosmos. When we look at things broadly this way, we see that the problems we humans have created may not be as great as problems other species have created, for which life found solutions. What could be more interesting, more exciting, than to be alive in the very age when we as a species have the opportunity to mature, to solve the adolescent problems we have caused ourselves and others?

O   O   O

One solution to human problems proposed in the name of ecology is that we should recognize technology as an inhuman disaster, an evil to be rooted out along with the science that produced it, so that we may go back to a simple, more natural life. After all, those humans who never invented technological languages and machinery, who never built big cities or hierarchical class societies, never got to adolescent crises of their own making. In fact, natives of the Amazon, New Guinea, the Australian outback and other places where people remained in settings relatively undisturbed by themselves, cannot be said to be immature in the sense that technological humans are. Though they belong to our very young species, they have never rejected the parental status or lessons of nature, never developed our kind of ego, simply learning deeply what it takes—and does not take—to live as one species within a mature and balanced ecosystem.

The rest of us—the vast majority of our species—must recognize that our development has taken a different path for reasons of its own and that no living system can reverse its evolutionary history. Our technological development is as natural as was our pre-technological infancy, and we cannot turn back. We could, however, move forward with more mature restraint and wisdom.

Fortunately, our industrial technology is already in transition from crude adolescent efforts to more mature sophistication. In developing our heavy industry and feeding it with raw materials, fuel, and human workers, we have devastated or polluted whole ecosystems, alienated ourselves from our natural origins, formed ourselves into mechanical societies living in concrete jungles. Yet the information age has already moved us into the next phase of living more lightly upon the Earth. We have begun replacing heavy industry with light, energy sources that can run out with those which won't, industrial cities with more distributed production networks connected through computers.

Many developing countries can now avoid the expensive and exhausting heavy-industry phase almost entirely by jumping directly into the age of electronics and information with the assistance of the most technologically advanced nations. This is already happening, but unfortunately it is happening in a context of profit motives and competition, often leaving beneficiaries worse off than before. They are forced to pay back assistance loans with heavy growing interest, to make political concessions and demonstrate loyalties. These development schemes continue to devastate ecosystems, as the World Bank has long admitted, and often help promote a climate of threatened, if not actual, hostility. Notorious are the post-colonial banana republics—single-crop economies that have created unstable ecosystems and dictatorships that keep their own people poor and hungry. Revolutions ensue; military might and the notorious 'disappearances' are used to crush them. Even where peace seems to reign, governments almost always work more for the interests of the rich and foreign investment industry than they do for the majority of their ever more impoverished people.

The building of dams to generate electricity, the burning and bulldozing of forests for monoculture crops or grazing cattle, the use of chemical fertilizers and pesticides, the manufacture of fuels and metals from fossil and ore deposits, are all financially profitable to those who own or rule the land. But they are so ecologically destructive that the life of ordinary people may become intolerable or non-viable.

Nature works not for profits but for balance, recycling everything. Humans cannot much longer run their profit-oriented growth economies at the expense of planetary economics, as Thomas Berry, the economist and philosopher priest, has warned. Paul Hawken tells us business will survive only if it adopts the recycling economy of nature. Business ethicist and professor William Frederick points out that nature's economy is about doing more with less—"the only way to survive, grow, develop and flourish." David Korten tells us we need "a story that gives meaning to life beyond an eternal competition for material acquisition and consumption." In fact, these voices are cropping up everywhere, making it evident that we recognize what has to be done.

If we are willing to see the problems we now face as those of a species on the brink of maturity, yet still in the fiercely competitive phase of belligerent adolescence, then we can learn as a species from our individual experience in maturation. Let us recall Mark Twain's now classic joke about the young man who comes home again after having struck out to make his way in the world. He is shocked yet pleasantly surprised to see how much his father has learned in the few years of his absence, how much they can at last agree on. The joke, of course, is that the son has changed and not the father—his own experience of the world having given him new perspective on his father as a wise and sensible man with something worthwhile to say.

This is exactly what can happen to us as a species. As a result of our own experience and our recognition of the environmental trouble we have caused, we can take a new look at the planet that gave us birth and we can begin seeing it in a new light. Through our brilliant science, our measuring and computing instruments, our space technology, we can see our planet as a whole living being that we have misunderstood and mistreated at our own expense. We begin to understand that while the planet has great experience and wisdom to teach us, our own lack of understanding and respect have led us to exploit it as though it existed for no other purpose. Only now do we understand that we have been recklessly destroying the parent planet on which we depend and from which we can learn a great deal about using our gift of conscious freedom more wisely.

O   O   O

In discussing the problems of free and conscious behavioral choice in earlier chapters, we spoke of our lack of innate rules for dividing land and resources fairly, for avoiding killing our kind, and for governing ourselves peacefully and cooperatively. We suggested that guidance in achieving these things by our own choice could be found in the organization and functioning of natural holarchies. We also recognized that turning our understanding into practice, instead of just reasoned ideas, requires a lot of responsible effort.

The Athenian democracy in ancient Greece was just such an effort, yet it seems to have collapsed at least as much from internal weakness—from the human reluctance to accept the hard responsibilities of freedom—as from external causes. Modern so-called democracies may work as well as they do only because they are much less demanding of their citizens than was ancient Athens, and this is important food for thought. In the United States today, compulsory taxation is the only requirement of citizenship, and compulsory taxation has never made a democracy.

Late in the nineteenth century, the great Russian writer Feodor Dostoevsky identified the problem of accepting responsibility for freedom as humanity's essential crisis. Dostoevsky presents the crisis of human freedom in a myth within his novel the *Brothers Karamazov*, by having one brother tell it to another. In this myth, Christ reappears in sixteenth-century Seville at the time when the Holy Christian Fathers, in that city alone, were burning as many as a hundred heretics a day at the stake in actual fact. A Grand Inquisitor condemns Christ to his second death—this time at the stake—for preaching freedom to mankind. The grounds given by the Inquisitor for this sentence is that "nothing has ever been more unendurable to man and to human society than freedom."

Men cannot bear, and so do not want, the responsibility of freedom, the Inquisitor claims, and the church has relieved them of that burden. Men will endure slavery for the sake of being fed and they will be happy only when their rebelliousness is turned to obedience, he insists, for they are sheep, preferring to believe they are free while actually doing as they are told by authorities who give them work, bread, rules to live by, and forgiveness for their sins.

Interestingly, Dostoevsky calls men "unfinished experimental creatures" in this passage, implying that they are immature as a species. Further, in his allegory, he deliberately pits two heroic images against each other in an analogy of personal adolescent crisis. Christ, on the one hand, represents the call to maturity—the acceptance of human responsibilities of free choice; the Inquisitor, on the other hand, represents retreat from maturity—the delegation of responsibility to an authority. Thus Dostoevsky portrayed his painful awareness of the human failure to practice responsible freedom.

Young people all over the world today reflect this dilemma in their personal lives, whether or not they are aware that it is as much a species dilemma as a personal one. Some struggle hard to develop and work in organizations that show all humanity how to accept the responsibility of freedom to end war, hunger, and ecological disaster. Others hurt so much from seeing the human failure to solve these problems that they fall into anarchy and depression, believing in no future at all and driving the youth crime and suicide rate to an unprecedented high. The rest—as Dostoevsky despairingly and accurately noted—do not concern themselves with such great problems but do what is asked of them to make their personal lives as comfortable as possible.

Perhaps most people search for comfortable security under some authority other than their own because we have been taught to fear failure if we fly in the face of tradition to exercise our own free choice. We have had the goal of perfection held up to us for thousands of years, and we fear failure and disapproval if we stray from whatever path we are told is the right one. Better to look to some authority for guidance, for the ideology, the formula, that will make us and our societies more perfect, than to risk acting on our own imperfect ideas.

For more than two millennia, ever since the ancient Greeks thought up the idea, we have been chasing after perfection. Now we are forced to wonder if we have not crippled ourselves in this chase after a *chimera*—a foolish and sometimes frightening fantasy. Let us consider it a happy discovery that the cosmos is not rigidly perfect as Plato thought, but an imperfect creative learning process much more as Anaximander saw it, with everything forming and re-forming in the never-ending process of making order from chaos.

Ever since Plato, western worldviews have held up the goal of making ourselves and our societies as perfect as God's creation—as perfect as well oiled machinery. Only now do we see that despite billions of years of experience, despite the marvelous integrity of life's patterns, things go wrong in Gaian nature, unbalancing the dance here or there. Yet in going wrong, they create pressure to reorganize the dance of life, to try new steps, or new combinations of old steps, and so the imperfection leads to progress.

The story of Earthlife is the story of improvisation wherever and whenever a species or ecosystem became unstable. Yet in nature there is never any break with the past—there is always continuity in the dance, even through extinctions. And the dance always works to produce a remarkable integrity and stability after periods of competitive strife. Nature teaches us that order can be maintained through change—even, when necessary, through disastrous change.

When, for example, the dinosaurs were killed off by severely changing conditions due to an accident, Earth's living systems continued to create new reptile, bird, and mammal species from the genes left in the small survivors of the disaster. We sometimes think of the great dinosaurs as unsuccessful species because they became extinct. But dinosaurs and their cousins flowered into a wonderful variety of species including the largest creatures the Earth has ever seen, and they flourished for nearly two hundred million years—forty times longer than our few million years as humans. Nor was their extinction their fault.

It is still up to us to prove that the human big-brain experiment is worth the risk, that freedom from innate rules—the conscious freedom to choose—will pay off in creativity that benefits, or at least does not harm, the whole Gaian system. If it doesn't pay off, we, too, will become extinct, more likely by our own doing than by outside forces as in the case of the dinosaurs. We would be wise to remind ourselves often that Gaia's dance will continue with or without us.

O   O   O

Nature, as we said before, is far more like a wonderfully resourceful artist than like a grand engineer, more like a mother juggling family needs, economics, and conflicts than like a coldly calculating geometer. In the improvised dance of nature toward order and balance, complexity unfolds, becomes chaotic or fragmented, is reorganized to new unity, then permits new complexity to unfold, new disorder to arise. This evolving system of life protects what is stable and works well, yet is ever open to change when instabilities arise, using change to create both new unity and new variety—variety that gives nature, among other things, the resilience to survive disasters.

Every species is different from all the others and every individual is a variation of that species' kind, just as in our bodies every organ is different from all the others and every individual cell a variation of that organ's kind. Machines can be mass-produced to be all alike, but nothing in nature is exactly like anything else.

It is Gaian wisdom to balance variety and use it creatively in forming highly stable ecosystems. The greater this variety is, the more stable the ecosystem is as a whole, as ecologists such as Eugene Odum and Edward Goldsmith have pointed out. We are also discovering that tampering with such systems by introducing a new species that has not been worked into the dance may disrupt it entirely—as in the case of the gypsy moth or Dutch elm disease.

This variety principle holds also for the gene pool of any species. We have learned by hard experience, for example, that our practice of 'perfecting' our food crops and domestic animals by breeding out their genetic variety, while breeding in the features we like, leaves them weak and subject to diseases or developmental anomalies. The first 'successfully' cloned sheep, Molly, for example, proved to age at ten times the normal rate for sheep. When we reduce variety by breeding a particular strain or cloning a single individual, by replacing natural ecosystems with monocultures on bulldozed land, we creating highly unstable and vulnerable situations.

Human variety, in our physical makeup as well as in our languages and cultures, ideas and lifestyles, is surely equally important to our healthy survival. Yet oddly, while we humans fight for our individual right to be different from others, not to be forced into the same social mold, we cause ourselves a good deal of trouble as a species by thinking there is something wrong with people who are different from us. We discriminate on the basis of color, culture, or belief—convincing ourselves there is something in difference to be hated, feared, ridiculed, or stamped out. Let us hope such prejudice will disappear as we learn more about nature and begin to respect, welcome, appreciate, and love our individual and cultural differences, using them like genetic variety to create new and fruitful combinations.

It has become obvious, for example, that a common human language is essential to communication and cooperation in the newly formed body of humanity, and English appears to be naturally evolving into that common language. But that is no reason to suppress any of the languages of different cultures. It is no problem at all for human children to learn several languages, and the variety of human languages represents a very important variety in human thought and worldviews. Just as we are foolish to breed natural variety out of domestic food plants and animals while killing off one wild species after another, we are foolish to eliminate variety in human language, culture, and thought.

Many natural languages have already become extinct as a result of foreign conquests. Conquerors have killed off conquered peoples and sometimes even punished survivors for speaking their native tongue, as was done in mission schools that children of native North and South American tribes were forced to attend. Just as we have begun to work at preserving endangered animal and plant species, we should be working to preserve endangered cultures and languages.

Over half the world's languages are already gone and it is estimated that half those remaining will be gone in one more generation. When we cut tropical forests, we destroy not only vital ecosystems but also the human cultures that are part of them, that cannot survive being transplanted to concrete jungles—cultures from which we can learn a great deal about living in harmonious balance with the rest of nature (see Chapter 19).

Nature's adaptive ability to change creatively without ever falling back into chaos surely suggests that we humans should give up the idea of finding the ideal economic and political organization or social structure. Our basic and natural task is the same as that of any other species—to balance individual good with collective species good, to be conservative with what is healthy in human society while radically changing what is not.

Nature teaches us that evolution depends on competition *and* cooperation, on independence *and* interdependence. Competition and independence are both important to individual survival, while cooperation and interdependence are both important to group, or social, or species survival. Individuals and their society or species are holons at two levels, or in two layers, of the same holarchy. We can see that these levels or layers must achieve mutual consistency by looking out for themselves *and* working out between themselves a balance of competition and cooperation, of dependence and interdependence.

If we work creatively to maintain this balance between ourselves as individuals and ourselves as societies—local, national, and worldwide—we will complete the evolution of a healthy body of humanity. If we look to our own individual bodies as a rough model for making it work, we might see that cooperative peace is a real option for nations with different languages and cultures that can make different contributions to the worldwide economy. There is no reason why individuals should not have the freedom to pursue their own interests and also contribute to their society. There is no reason why all should not be well fed and cared for in an equitable system of work and income.

O   O   O

From the Gaian perspective, solutions to the great human problems of war, overpopulation, and hunger are far simpler than from any other perspective. But simplicity does not mean ease. We *could* solve these problems by shifting our worldview from one of international competition to one of international cooperation, with the goal of producing a healthy body of humanity. But worldviews, as we said earlier, do not change easily.

The present perspective of the powerful nations, banks and multi-national corporations—those powers capable of quickly transforming humanity into a healthy body—is not a Gaian perspective. To attain the Gaian perspective, their leaders must change their world-view and the behavior based on it. As it is, these leaders are uneasy at predictions of doomsday—at suggestions that their course is suicidal for all humanity— but they do not yet believe in a healthy alternative that would be as good for themselves as for all humanity.

The Cold War decades brought the world to the point where arms manufacture cost four times as much as it would cost to retire the developing nations' debt, provide worldwide clean, safe energy, housing, health care and clean water, stabilize populations, eliminate starvation and malnutrition, prevent ozone depletion, acid rain, deforestation and soil erosion, according to an analysis by the World Game Institute. What could more graphically show the bizarre distortion of our human endeavors? During the Gulf War, the five peacekeeping nations sent by the UN were the very ones that provided arms to both sides in the conflict!

It is no easier to get the citizens of our modern democratic or communist societies to take on the informed and responsible task of running things, instead of submitting to the oligarchic rule of the few, than it was to get ancient Athenian citizens to do so. But trends toward more networks in place of top-down authoritarian structures are emerging in new organizations and in various existing industries, services, and other enterprises as they decentralize their management and make it more concentric than hierarchical. But nowhere is this trend as much in evidence as in the Internet, as we will see in more detail later. As more people come to understand and adopt a Gaian perspective, this trend will surely grow.

Just as individuals may grow out of adolescence without consciously assisting their own process of maturation, a cooperative body of humanity may evolve without our conscious intent—just by the force of evolutionary change that is already under way. But the process will surely continue with less turmoil and suffering if we stop opposing it and consciously, actively assist it. Would it not be healthier for us to give up our dangerous competition for the greatest financial profits and work together at creating an imperfect but ecologically sensible economics and politics—a system that works with its variety to give everyone the opportunity for a healthy life, just as our bodies give that opportunity to each of their cells?

In recognizing our planet as an experienced living system with a good deal of accumulated wisdom to teach us, we gain the perspective to see how we might apply some of that wisdom to our own human problems. All over nature, throughout the Gaian life system—right under our noses, so to speak, and all around us—we find the clues to making our own human affairs more organic and ethical, more creative and wise, as the earliest philosophers believed we would.

Let us continue gaining perspective by seeing that great problems can be the very challenges needed to push evolution along into new creativity. We saw that the oxygen crisis of billions of years ago became an opportunity for a new way of life and new forms of living creatures. The environmental crisis that caused the mass extinction of dinosaurs provided the opportunity for our own mammalian evolution. Both the bacterial oxygen crisis and the dinosaurs' extinction crisis were more severe from a Gaian perspective then any trouble we humans have caused *so far*. Yet we have already initiated another extinction and it increasingly appears to be a life or death crisis for our own species. Our increase of the greenhouse gases alone may force Gaia to regain stability at an average temperature beyond human tolerance. Ozone holes, nuclear holocausts, poisoned air and waters, chemically depleted soils, epidemics caused by microbial defenses against our war-on-life antibiotics—we seem to be capable of inventing a remarkable number of potentially species-suicide weapons.

Why not put our cultural and ethnic differences and imperfections to creative use in dialoguing about a healthier future? Why not see the crises of our making as incentives to move forward in new ways. Gaian creativity might have come to an end without problems, challenges, opportunities for creating solutions leading to new species and ways of life. If we follow nature's lead, we will make mistakes, juggle things about, find solutions, generate new problems without guilt—and on the whole, we will find our mutual consistency with other natural organisms and with each other.

Let us also remember that if we continue on our current path, our planet may be better off without us. Our species demise—by suicide or extinction—might actually promote Gaian health.

Yet we are potentially as creative as the whole Gaian system we belong to. If we find ourselves in an adolescent crisis of our own making, that is no reason for us to give up in despair. It should, instead, urge us to face ourselves, swallow our foolish pride, adopt a little humility, a wider perspective, and gain mature humanity in the best sense of this word we have coined for ourselves.

The wider perspective many humans are waking to now is the perspective that we are not humans capable of having spiritual experiences, but spirits having human experiences. This perspective was until recently found only among religious people, but with new discoveries in physics we talked about earlier—such as evidence of cosmic consciousness and intelligence, and the non-locality of a completely interwoven universe in which everything affects everything else at any 'distance'—scientists and other lay people are joining their ranks. This worldview connects especially easily with Gaian science and philosophy in Buddhism, which is enjoying great outreach in the West.

Another past province of religion now broadening its base is ethics, since science, in its love affair with objectivity, divorced itself from such concerns. Now we find there is no possibility of cold objectivity in a participatory and interwoven universe. Perhaps we can even find ethics built into nature itself.

18

Ecological Ethics

Let us begin this search for natural ethics by seeing that there is reason enough apart from our youth to adopt a little species humility. We have seen that no species can evolve apart from its co-evolution with all other species—meaning that *all* have played their role in *our* evolution. We could not have evolved by ourselves. If we look at co-evolving living systems through eyes other than our own, we will quickly see that we have no more reason to consider ourselves a supreme form of life than have others.

Recall that mitochondria and chloroplasts, descended from ancient bacteria, make up half the weight of all plant and animal cells, causing Lewis Thomas to call us giant taxis for bacteria to get around in. Joking aside, the world from a bacterial point of view is indeed arranged nicely for bacterial survival. They live not only in their colonies and fabulous cities, but can and do live in—or buzz in and out of—all other forms of life, feeding off the living and the dead, passing around bits of DNA information in their WorldWideWeb.

They exist in vastly greater numbers than any other kind of living creature, and there is virtually no place on Earth—from the depths of the sea to the highest mountain and the atmosphere itself, from the hottest springs to the coldest glaciers, from the surfaces of other creatures to the depths of their guts—that is not teeming with them. They spread over Earth's entire surface and evolved even its geological features—including the atmosphere and entire continental shelves and veins of minerals, transporting them about in quantity, forming the ore veins, such as the copper and uranium we mine today. All this they did by themselves for half of Earth's life, while even today they maintain a good deal of its functioning and balance.

Bacteria are responsible for forming the larger cells from which all other life kingdoms are constituted. Further, bacteria are the *only* creatures that could survive without all the others. Why should bacteria not think—if they could think—that the world is all theirs?

Then, take the fungi—a kingdom of life in themselves. They, too, are spread out almost everywhere, and though most are too small or fine-webbed for us to see, some are so extensive underground that we know them to be the largest creatures on Earth! Every plant of Earth has funguses twined in its roots, bringing it supplies in return for ready-made food. Funguses live on animals as well as on plants. From their point of view, all nature would seem to have been created as their dinner table.

Animals might well look this way upon plants—as though plants had been created especially to feed and serve them. After all, animals eat plants to burn out of them with oxygen the energy they need—and that oxygen was made and put into the atmosphere by the plants themselves, as if to supply animals with breath as well as with food. Animals make use of plants even for their drinking water, lapping or sucking dew, drinking rainwater that was first pumped into the sky through the roots and stems and leaves of plants. Animals also use plants for shelter, making their homes in seaweed and among the branches and roots of land plants from the tiniest club mosses to the tallest trees.

Surely animals could not be blamed for believing that plants had evolved just to provide them with food, oxygen, water and shelter. But what if we shift our perspective to that of plants?

From their own point of view, plants might very well think that animals were created to provide for *them*. Plants—that vast range of photosynthesizers from little more than fancy bacterial colonies to great banyan trees, each a forest in itself, have considerable reason to see themselves as superior creatures. Recall that animals evolved from creatures that *lost* their chloroplasts and thus had to spend their lives chasing after food. Plants need not run about chasing after food, but can sit right where they are, easily making their own food and energy from sunlight and soil chemicals provided by bacteria, funguses, worms, insects, and other animals. Animals also provide the carbon dioxide they use in making energy. Insects carry pollen from plant to plant, making it easy for plants to reproduce without running about for that purpose, either.

When plants have made their seeds, animals continue to work for them, carrying the seeds about in their feathers and fur. Birds and grazing mammals also eat the plants' fruits and digest them. The animals thus moisten the plants' seeds and wrap them in packets of rich fertilizer, then scatter them in new places to grow. Animals, in fact, do all the running around for plants while plants sit smugly being served all their lives wherever they first took root.

And so all these forms of Gaian life—bacteria, fungi, plants and animals—could find reason to see themselves as superior to the others. Even rock, for that matter, could see the whole world as nothing more than its own dance, its endless transformation into living creatures and back into rock. Try for yourself the exercise of looking out over your world and seeing all of it—the landscape, the sea and sky, the creatures, yourself and your fellow humans, their airplanes, their cities, the furniture in your house, this book in your hands—all as no more and no less than rock rearranged.

The continents of the Earth are still on the move. Ever since the great single continent, Pangaea broke apart, they have ridden their tectonic plates slowly over the softer mantle beneath them. Africa and South America only separated around the time the dinosaurs disappeared, some sixty million years ago, as we saw. By moving apart, rock separated its life forms, cutting species members off from one another so they were forced into different lines of evolution. From rock's point of view, it directs even the course of evolution through its motion.

Rock might think it had recruited bacteria and protists into the work of rearranging its minerals. Protists, for example, have long been engaged in that endless work of moving calcium and silica about in huge quantities by building it into their shells and depositing it on the sea floor, which later thrusts itself up to become land. In the great carboniferous forests, plants were recruited to begin *their* job of burying carbon underground to become coal and oil. And so the Earth's rocky crust, itself formerly stardust, has reason to see the bacteria, protists, plants, and animals created in its metabolic dance as its own inventions, meant to serve its needs.

We humans, from all these perspectives, would be considered late-comers—an upstart species coming in to upset the whole dance by killing off or endangering others as we make war on all five king-doms including our own, as well as the crustal formations. We burn and cut forests, dam up and choke off rivers, create deserts, poison water, air, and soil. And in our unprecedented egotism, we behave this way while declaring ourselves the pinnacle of evolution!

Yet we alone are capable of holding a truly broad worldview that represents the whole of nature and includes all possible points of view in addition to our own, as we just saw. We can—and we must—gain enough perspective to see ourselves as one part of a much greater living system, or being, and learn to act accordingly. The body of humanity we have described in its present evolution is a new kind of body and, at the same time, an organ within the Gaian body—the latest organ to evolve within it, one that is only now being tested to see whether it can function. Humanity has woefully little experiential intelligence or wisdom, yet it must evolve by its own free choice among alternative conscious ideas, decisions, and practices.

O   O   O

Human ideas, concepts or pieces of information that become known and passed on by large numbers of people have come to be called *memes*. This name is intended to show a parallel with genes—in the sense that memes can spread through human popu-lations in patterns that influence their social evolution as genes are passed on in patterns influencing biological evolution. It is from memes—particular ideas about ourselves and our world—that we construct the worldviews that shape our societies.

What memes are now determining the formation and nature of the new body of humanity? The memes from which we have built our dominant culture worldview include, for example, ideas of ourselves as divided into competitive nations—nations or blocs of nations competing for resources and power much as the ancient bacteria competed with one another before they formed a common protist nucleus and thus a protist identity.

Such ideas of separate nations in competition with one another must be transformed into ideas of cooperation among varied nations as organs in a single body. Blocs of nations such as NATO, APEC and the EEC have already been formed and function like organ systems that can carry out different tasks such as defense, economics and monetary systems. These systems, however, are independent of the United Nations and in some ways at cross-purposes with it. We still lack the overall perspective from which to form a common plan, though our evolution to date has prepared us to make it.

If we transform the memes composing our worldview to a common scheme of voluntary cooperation—a self-creative, autopoietic, body of humanity—we will abandon our old ideas of the machinery of society and find new organic ways of reorganizing ourselves.

Bioregionalists, such as Van Andruss, Christopher and Judith Plant, have proposed, for example, that naturally bounded ecological areas such as watersheds make more sense as economic-political units than do our present states or provinces, with their arbitrarily drawn boundaries. The inhabitants of such areas would have natural interests in common. The natural boundaries of ecosystemic regions contained original human societies, which recognized their dependence on nature and did not yet see it as territory to be carved up arbitrarily. But most of us today do not *think* bioregionally, because we lack the concept as a cultural meme.

Unfortunately, we have lost some of the most important memes generated in human history. The massive upheaval of human society that began six thousand years ago, initiating an empire-building era lasting to the present—was so thorough in promoting dominance and aggression over partnership and peace that we came to see such aggression as our natural heritage. Can we recover the memes of civilized equality and peaceful sharing of wealth that seem to have guided settled human societies during the preceding thirty or so thousand years?

Bioregionalism proposes that the inhabitants of an ecosystem, such as a watershed area, assess the natural species living there and the region's capacity for supporting them as well as the human occupants. The humans would then work out the rebuilding of community in harmony with its ecosystem, aiming at satisfying human needs locally as much as possible, within sound ecological constraints and importing only what is necessary. Bioregionalism could be a working model for the whole body of humanity, with careful urbanization and harmonious agreements on regional production and trade across regions, especially if combined with Hawkens' proposal that we emulate nature by eliminating the concept of waste, so that everything we produce is consumable or recyclable.

However we draw boundaries and organize ourselves, our new body of humanity must be flexible enough to evolve through still further stages. We can be sure that it will always be imperfect by the old mechanical standards. Whatever social forms it will eventually take, making the body of humanity into a healthy, functioning holon within the Gaian holarchy is the greatest task human consciousness has yet faced.

O   O   O

The Gaian system as a whole is part of the larger conscious cosmos. It is an intelligent system that knows itself, as reflected in our bodies' self knowledge—all parts interconnected by all manner of communications. But it does not have our unique human consciousness, with its ability to see Gaian history linearly, to abstract patterns from its great complexity and to link past to future by planning. It does not seek to explain itself to itself, or make models of itself to use in deciding what to do next. Yet the Gaian system functions intelligently and wisely, learning throughout its billions of years of experience.

Human consciousness is the newest Gaian experiment, and in Gaian wisdom little has been entrusted to it as yet. Our bodies continue to manage themselves mostly without our conscious help or interference, and this is most fortunate, given the complexity of their functions. Our conscious scientific minds—no matter how impressed we are with our ever growing knowledge—are very far from understanding even a single cell or organ well enough to manage its ordinary daily affairs. Lewis Thomas recognized this in saying he would rather have to fly the most complex jet without training than to try to manage his liver for a day. Even a single cell is so complex we are still trying to understand it, and still making new discoveries about it.

It should come as no great surprise that the freedom of conscious decision making gives us a good deal of anxiety. We look around us and see other species functioning on the whole the way our bodies do, untroubled by questions of whether what they are doing is right or wrong, good or bad. Yet we are stuck with choice-making conscious minds that are an experimental substitute for the innate evolutionary knowing of other species, and we must use those minds as best we can to decide how to behave.

It is because of this unprecedented degree of choice that we humans alone must ask ourselves what is right or wrong, good or bad for us to do—personally, socially, and as a species. Modern science, however, has refused to concern itself with such questions—on the grounds that they are ethical questions, which it sees as the domain of religion, not science.

When sixth-century B.C. philosophers suggested that ethical questions can be answered by looking to nature, religion and science had not yet been separated, both being aspects of the same search for orientation and guidance. Only much later, when modern scientists separated them, did ethics become the domain of religion while scientists insisted on their ethical neutrality, their freedom from values.

If the ancients were right—as this book holds—that nature is a source of guidance for human behavior, then surely science, as the study of nature, *should* concern itself with ethics—with showing us what is wise or not wise to do in our relationship to one another and to the rest of nature.

We have seen historically how we strayed from this path. Since the time of those early philosophers we have come to see our own consciousness as an ego, or I, that could be set apart from nature—an objective 'eye,' viewing it, making theories about it, evolving a worldview to explain it. We developed a technology to help us view it through lenses and exploit it through machinery. Yet for all its early religious roots and in spite of its technological success, the mechanistic worldview that brought us to the present left us facing enormous problems without any scientific ethics to guide us in solving them.

We even thought we did not want such guidance, having become weary and leery of the mere mention of ethics. History has shown that ethics—as traditionally defined and promoted by religious authorities—has been used to make people obedient servants of those in power more often than it has been used for their own good. (Recall how the Greek definition of virtue as excellence was changed to mean obedience.) Still, when we are not handed our ethics, as was the Grand Inquisitor's flock, we are more than ever in need of guidance to overcome the anxiety that comes with freedom. We can hardly expect ourselves to take the responsibility for our actions until we have some way of judging what actions are right or wrong, good or bad.

The word religion comes from the Latin *re-ligio*, which means 'reconnect.' Religion is a way of reconnecting ourselves to our origins, to our source. These origins are seen in religions as the creative acts of one or more deities, connection with which—through prayer or priesthoods—gives us guidance. But science, too, concerns itself with our origins, and thus reconnects us with source through a story of natural evolution. Why *not* find some kind of guidance in *these* connections? Is it not likely that nature built into us this need to reconnect because our survival depends on maintaining connections with our origins, including understanding our relation to our co-evolving environment or ecosystem so that we may play a balanced role within it? Having given us free choice, would nature not also have built in some guidance for making our choices?

We have already discussed the ethical guidance to be found in the organization of our own bodies, even in the evolution of our cells. But what of the popular impression that nothing could be more *unethical* than nature? To conclude that nature is cruel and insensitive, we have only to think of a panther attacking and killing a baby gazelle, an owl pouncing on a mouse, a praying mantis biting off her own mate's head, a wasp laying her eggs inside a live caterpillar, which her children will eat from the inside out. Many species defend themselves viciously by human standards—plants with deadly poisons, spiky thorns, glassy stingers; butterflies eating poison plants that kill birds eating *them*; sea urchins leaving humans who step on them in pain for weeks; lions tearing up their hunters as well as their prey. Is all this not wanton, unethical cruelty?

Such arguments cannot be dismissed lightly, but let us remember that we make them from a human point of view. Indeed they reflect healthy human sensitivities which we shall discuss shortly. But to understand the Gaian system, we must see it also from the perspective in which life rearranges our planet's rocky crust into a multitude of species, all parts of a single whole in which they are necessarily recycled.

Right from the start, the first bacteria—living packets of enzyme-driven giant molecules—could build themselves only by consuming and using smaller molecules that had also been part of Earth's crust. Many of these molecules later were built back into rock; others became parts of new bacteria. Without such recycling, where would Gaia be? The world would have filled with bacteria that would simply have died when supplies ran out—as the bubblers almost did—and that would have been the end of it. Earth would have ended up as lifeless as Venus and Mars are today.

Recycling is the secret of life's endless creativity, and we humans are just beginning to understand the problems our own species has caused by using things up *without* recycling. As supplies of original molecules ran out, living holons were forced to use the manufactured molecules of other living holons. As we saw, for example, bacteria began consuming one another. Much later, animals had to evolve the equipment for chasing after their food rather than sitting in one spot making it like plants, which could make food from sunlight and local chemicals recycled by other species.

Thus large living holons evolved and maintained themselves by incorporating smaller holons and thus recycling them. What disturbs us is the fact that their food is often alive and must be killed to be eaten. But why are we so disturbed by these things which are in our own nature to do? Do we not protect ourselves against attack with the best weapons we know how to make? Are we not hunters and killers ourselves? Even vegetarians tear plants limb from limb, boil them to death, or crunch them up raw with grinding teeth. We cannot get along without feeding on other living things.

Is it the infliction of unnecessary pain and suffering that bothers us? Then let us note that our modern means of producing poultry and livestock, not to mention our use of research animals, is more cruel by our own standards than other species' means of killing. Our methods involve the lifelong torture of being imprisoned in extremely limiting boxes or cages or feed lots, with no access to ecosystems and conditions so unhealthy they must constantly be fed medicines. Animals chase and kill their free-living prey quickly. Further, there is good evidence that bodies which have evolved the capacity to feel pain as a trouble signal know when they can no longer protect themselves and turn off their pain system so they will not suffer needlessly. This seems to be so for the wounded mouse in the claws of a cat as much as for the human soldier wounded in a battle with his fellow man for far less justifiable reasons.

Plants use recycled supplies in ways that don't bother us. Though the molecules they take in through their roots may have been part of large creatures, they are taken in only when busy bacteria have decayed the creatures. The way bacteria eat doesn't bother us either, as they and their habits are quite invisible to us.

Rather than justifying our cruelties by accusing nature of cruelty, we should look at our own ability for compassion and do the least damage we can, as Buddhists urge.

O　　　　O　　　　O

This is the key to natural ethics—that the self-interest of *every* level or layer in a holarchy is the best possible strategy, for *only* by means of that strategy can mutual consistency work itself out among all levels. We've already seen how this works, for example in our bodies, but let's look at the principle itself more closely. *Mutual* means shared, and *consistency* is harmony or agreement—shared harmony is what we have called ecological balance. When every holon in the holarchy of an ecological system looks out for itself, a shared balance, or mutual consistency, results. A species holon, for instance, keeps itself healthy by producing a variety of offspring in competitive numbers, of which the healthiest are most likely to survive and become the parents of its next generation. But the species holon needs help in this selection—help from the larger holon, which is its ecosystem and which is made of other species.

Among animals, it is clear that hunters are most likely to catch the weakest members of a prey species. A hunting species thus actually helps its prey species to stay healthy by weeding out its weakest members, while the prey species helps the predators by keeping them fed. Mutual consistency often involves mutual benefit.

Most species live on a very limited diet; many will eat only a single kind of food—koalas, for example, eat only eucalyptus leaves and anteaters consume only ants. If a species eats too much of its food species, it will lose its own members to starvation. This gives the food species a chance to recover, and when it does, more of the eating species will live again. In this way species rebalance their imbalances and restore one another's health. Mutual consistency implies a continually dynamic balancing process.

We saw other examples of mutual consistency between individuals and their species in the territoriality and social structures discussed earlier. Individuals fighting for their own territory, for example, do so in such a way that the whole species benefits from protection against overcrowding and inadequate resources.

Nature tests the evolving patterns of species and their ecosystems against one another to see that they are in, or are able to restore, balance—that they share harmony, or that they are mutually consistent. Whatever proves unable to gain consistency with all else around it cannot survive. This testing is seen as the progression from a new ecosystem, in which a few species compete for territory, to a mature ecosystem in which many species exist and demonstrate their mutual consistency.

Can any of us think of a better way for life as a whole to keep itself alive and in good health? The system is worked out so that every part looks out for itself without taking more than it needs and *in doing so* contributes to the welfare of the whole! Every part thus finds its dynamic balance with every other part, working out mutual consistency in such a way that the whole Gaian system works as a single healthy being—every part, that is, except the experimental new human species, which does take more than it needs, wantonly destroying whole other species and ecosystems in the process, killing and starving large numbers of its own species, all the while accusing the rest of nature of cruelty.

What shall we make of this human sensitivity that lives in us side by side with our own cruelty? What of this pity that we profess to feel for other creatures as we torture and slaughter them? Both aspects of this human portrait are unique; no other species demonstrates *either* the wanton slaughter of, *or* the pity for, other species. Could these feelings of pity be made to serve some useful purpose in the dance of life, as most things surviving in nature seem to do? Could the feelings serve to awaken us to our own reckless cruelty and push us toward mutual consistency with other species?

Compassionate concern for others, as we saw, evolved in mammals along with emotions, behavioral choice and the birth of live young that needed care and teaching. A great deal of human behavior is guided by feelings, for better as well as for worse. As we humans have the freest behavior, the greatest choice, we might expect ourselves to have the strongest and most varied feelings. And so it seems to be.

When our feelings take over completely, we lose our ability to think about the possibilities and consequences of choice. On the other hand, when we use only our ability to think, we become cold and calculating in a way we think of as mechanical, or 'inhuman.' Like everything else in nature, human thought and feeling are ever in need of balancing.

But let's get back to the particular human feeling of horror at nature's cruelty.

When Darwin announced that competition over inadequate resources was the sole driving force of evolution, his theory, as we saw, was quickly used to excuse cruelty in the human world. If all nature was a bloody battle 'red in tooth and claw,' then why should humans be an exception? The competitive exploitation of resources and labor by the rich as they built an industrial world was thus justified on the grounds that it was natural. Even now there are sociobiologists who believe that aggression is our innate and therefore unchangeable animal heritage.

Another problem we have with ethics is that anthropologists, who study various human cultures, have told us that ethics is really no more than a set of behavioral codes specific to individual ethnic groups. What is ethical in one culture is unethical in another, so there is no point trying to find common human ethics.

Yet, common human ethics is what we now need more than anything else—ethics to guide us in our behavior toward one another and toward other species in the natural world to which we all belong. Our basis for such ethics becomes very different when we no longer see nature as just a bloody battleground for competitive struggles over limited resources.

Competition is merely one aspect of nature's creative organization into mutually consistent holons within holarchies—a midstage in the unity—>unity cycle [illustrated in Chapter 2]. *Young immature species* are the ones that grab as much territory and resources as they can, multiplying as fast as they can. But the process of negotiations with other species matures them, thus maturing entire ecosystems. Rainforests that have evolved over millions of years are a good example. No species is in charge—the system's leadership is distributed among all species, all knowing their part in the dance, all cooperating in mutual consistency. The best life insurance for any of the species is in giving off quality products useful to other species—what a lesson for us new and immature humans!

What we see clearly in such mature ecosystems is that every holon's health depends on the health of the larger holons in which it is embedded. Thus every holon, in looking out for itself, must also cooperate with other holons to help look out for their larger holon's interests.

This, as we said, is the heart of ecological ethics—the self-interest of *every* holon, whether a cell, a body, a society, a species, an ecosystem, or a whole living planet. All must be balanced in the mutual consistency of the whole and all its parts. *Self-interest is bad only when not tempered by the self-interest of community.*

For us this means recognizing how much we affect the living planet of which we are part and on which our continued existence depends. To truly look out for our own interests *requires* that we know the interests of our whole environment, which means our whole living planet. Our free choices, in order to serve our own long-range interests, must serve those of other species as well, for natural ethical behavior is that which contributes to the health of the whole Gaian system.

Our history has brought us to the shortsighted adolescent selfishness of warfare, hatred, distrust, and reckless destruction of our own environment. We have long-standing habits of believing that all nature is human property, and so we take land and resources from one another for reasons of profit. It is high time for us to realize that maximizing individual profits minimizes human social stability and welfare, while maximizing common profits destroys our natural life-support system. If we want to survive as a species we must learn to change our ideas and our lifestyles to live in a balanced recycling economy like the rest of nature.

In fact, it is high time to realize that all our old habits and vested interests, even if they form our individual and national identity, must be fundamentally changed. The changes required are deeper and more far-reaching than any revolutionary leader has ever demanded or even dreamed of demanding. And yet we can make those changes peacefully, and everyone can win.

One of the ways we are learning this in affluent countries is through the voluntary simplicity movement launched by pioneers such as Duane Elgin and Vicki Robin. Though a few of us may want to go back to the land building our own houses and chopping our own firewood, that is not what simplicity dictates. We need to discover— or rediscover—*elegant* simplicity, such as the Japanese and Balinese cultures have role-modeled. There can be many wonderful ways to do this in different cultures. Wouldn't it be wonderful to get rid of excess possessions requiring our time and attention, of junk that crowds our existence? How *freeing* to live lightly on the Earth, not only as volunteer individuals, but as a human species.

Such deep changes in humanity cannot be made at the point of a gun or by other kinds of force. They must be made voluntarily and that is perhaps more difficult. The profit motive is so ingrained in western society, for example, that scientists have actually criticized nature on the grounds of unprofitable inefficiency, pointing out that photosynthesizing plants use only a small fraction of the energy available in sunlight. Can such people learn to appreciate the fact that plants extract exactly as much energy as they need for themselves and to keep their environment's careful balance of energy exchange?

O   O   O

Our age-old religious quest for reconnection with origins has been the search not only for our origins, but for our Creator as an inspirational source of guidance and security that would lead us to a better life. In the early childhood of human civilization we imaged this source in sacred nature itself, symbolized by the Great Mother. Then we shifted our attention and loyalty to a Father God, casting him in human image, making him a mathematician when we invented mathematics and an engineer when we invented machinery. In our adolescent cheek, science rejected the father God, believing there was nothing greater or more intelligent in the entire universe than ourselves.

Now, on the brink of maturity, we can see that our earliest intuitions were most valid. The source of our creation is indeed an inspirational being far greater and wiser than ourselves—a Gaian being that has nurtured us and can guide us to a better way of life. Gaia, our living Earth, is not a perfect superhuman parent, but the imperfect yet wonderfully resourceful planet of which we are one part, and which is itself part of a far greater being, a Conscious Cosmos. Have we the maturity to trust and heed these sources of our being for their guidance?

19

The Indigenous Way

We have repeatedly observed that the dominant western culture of humanity, imposed on much of the world, behaves immaturely from an evolutionary perspective. We also suggested it had something to learn from the organization and evolution of ecosystems, as well as from some of the non-technological indigenous and traditional cultures that have survived the colonial process and the more recent impetus to modernization. In this chapter we will explore the worldviews and knowledge of indigenous peoples to see why cooperation between indigenous and industrial humanity is so important at this critical time in our evolution as the body of humanity.

Indigenous cultures are generally held to be non-industrial cultures with ancient roots in their land, though some have been migratory and others forcibly displaced. Their cultures range from very simple material lifestyles to extensive historical urban/rural systems such as Maya, Inca and Aztec. For all their great diversity, we will see that they do hold some common elements of worldview and values that unite them with each other and distinguish them from modern or post-modern industrial cultures, which are also diverse, yet united by their basic worldview and values.

In today's world, there are very few even relatively intact indigenous cultures. Yet we do have indigenous people to whom traditional knowledge and ways have been passed on and who live by this knowledge. This knowledge represents a relationship with the rest of our living planet that has been essentially rejected by industrial culture, yet is very relevant to our healthy future.

Let us begin with the historical perspectives of two indigenous cultures that have ancient teachings concerning their relationship with industrial cultures: The Hopi Indians of North America and the Kogi of South America.

The Hopi have an ancient prophecy predicting our present and future, reported, among other places, in Rudolf Kaiser's book *Voice of the Great Spirit*. Part of this prophecy tells the history of the Red and White Brothers, sons of the Earth Mother and the Great Spirit who gave them different missions. The Red Brother was to stay at home and keep the land in sacred trust while the White Brother went abroad to record things and make inventions.

One day the White Brother was to return and share his inventions in a spirit of respect for the wisdom his Red Brother had gained. It was told that his inventions would include cobwebs through which people could speak to each other from house to house across mountains, even with all doors and windows closed. There would be carriages crossing the sky on invisible roads, and eventually a gourd of ashes that when dropped would scorch the earth, burning everything, even the fishes in the sea.

If the White Brother's ego grew so large in making these inventions that he would not listen to the wisdom of the Red Brother, he would bring this world to an end in the Great Purification of nature. Only a few would survive to bring forth the next world in which there would again be abundance and harmony. (It should be mentioned that not all Hopi approve of having any part of this prophecy in print; the author apologizes to anyone who may be offended by this citation of other written sources.)

The Kogi Indians of Colombia have a similar historical scenario in their creation story, told as part of the BBC film made by Alan Ereira, called *Message from the Heart of the World: the Elder Brother's Warning*. According to the Kogi, the Great Mother Aluna is the primeval waters and the source of all creation. Even before creating worlds, she lived through all possibilities for all worlds and all times through great mental anguish. For this she is known as Memory and Possibility. The eight worlds she created previous to this one were not peopled, but in this ninth world she put humans, including Elder and Younger Brothers.

From the beginning, Younger Brother caused so much trouble that eventually he was given knowledge of technology and sent far, far away across the waters. Five hundred years ago, the Kogi say, he found his way back across the waters and he has been causing trouble again ever since. If he does not listen to the Kogi, who see themselves as Elder Brother, and stop destroying the Mother, stop digging out her heart with his mining and cutting up her liver with his deforestation, he will bring this world to an end.

From the Hopi and Kogi perspectives, we see that the white or younger brother of their ancient stories dominates present human existence. He is industrial man as we have seen him in earlier chapters, creating a technological society founded on a mechanical worldview and scientific discovery. We have seen that his technological way of life, for all its benefits, has brought us to the brink of disaster. In this chapter we will see that his ways stand in sharp contrast to many indigenous and traditional peoples' worldviews, value systems and lifestyles which are only now beginning to be recognized as valid in their own right and possibly critical for our very survival as a species.

The Hopi, with the help of many friends, made forty-five years of effort trying to tell their prophecy orally in the United Nations, succeeding at last in 1993, at the beginning of the UN Year of Indigenous Peoples. Their prophecy does not suggest we would be better off without industrial society. It does suggest that the wisdom and knowledge of indigenous peoples must provide the context in which we make, use and dispose of industrial goods if we are to survive. This view of things from their perspective is consistent with our own growing understanding of the need for ecologically sustainable development, as discussed in the next chapter.

It is important to understand why the UN resisted Hopi efforts to give their message for so long. Only if indigenous nations were granted sovereignty and recognized as nations could they have UN member status. In the meantime, the UN struggles to define their status and rights, given that they exist within member nations who do not wish to grant them this sovereignty.

Historically, the European colonial "White/Younger Brother" had seized the lands of the Hopi, the Kogi and most other indigenous cultures around the world on grounds dating back to a Papal Bull of 1493 stating that infidels had no land rights, while Christians did. Indigenous peoples were defined as part of the 'brute nature' the Europeans were to conquer and subdue; thus their territories were reduced to reservations within the boundaries of United Nations' member nations. Since this colonial process began, the Euro-American culture has perpetuated the dogma that indigenous people were backward, ignorant and impoverished before the white man's benign intervention. Jerry Mander and Chief Oren Lyons have both documented this unfair and brutal historical process with respect to indigenous North Americans.

Technological culture defines itself as progressive and non-technological cultures as backward and ignorant, thus taking the stance: What advice could they possibly give us? Only now, when we begin to understand how essential diversity is to the very survival of living systems, do we open ourselves to respect for different worldviews and the choice of different lifestyles.

The Hopi and the Kogi are only two among many indigenous cultures that have ancient prophecies of man's destruction of nature as well as present evaluations of our global crisis. These two in particular foretold not only nature's destruction at this time, but specifically identified, as we saw above, the inventive, technological branch of humanity as responsible because it fails to heed the sacred Earth knowledge and wisdom so vital to indigenous peoples. Yet neither the Hopi nor the Kogi tell us that technology is bad in itself, that we should abandon it and "go back to nature," living as they did. Both Hopi and Kogi validate technology as an important aspect of humanity, simply warning us that it must be brought into harmony with the sacred natural world.

How did these indigenous peoples know the crisis technology would bring on? Why is it that the science on which our technological world is based—the science which so prides itself on its ability to predict—failed to predict its own consequences while indigenous cultures saw where it would lead?

The failure of industrial society's scientists to predict the consequences of their technology is directly related to the mechanical/materialist worldview in which that technology was developed—a worldview fundamentally different from the organic worldviews of indigenous peoples. In the worldview shared by indigenous peoples everywhere, despite many differences in its formulation, the universe, nature, is alive and sacred. All beings within it are related and interdependent: the stars, the rocks, the waters, the winds, the creatures, the people, the spirits and so on.

The human role is to hold nature sacred and to live in a balanced way within it, to give back as much as is taken while pursuing social and spiritual development. There is no concept of waste and no waste accumulation. In many cases there is deliberate avoidance of material accumulation of any kind. The Northwest American potlatch ceremonies were designed as *giveaways* to distribute accumulated goods so that no one would be burdened by owning too many of them.

The scientific worldview of the conquering industrializing cultures held that the universe is fundamentally lifeless, with life happening by accident on the surface of this planet. In this view, which we have deeply explored, the role of science is to study nature objectively—as though from outside—and reduce its machinery to basic parts in order to understand it. One of the basic laws of nature in this view is the law of entropy discussed in Chapter 14, a law stating that everything in nature is running down, a law of unsustainability. The purpose of science is to gain control over nature by exploiting it for human purposes—converting it to food production and the manufacture of goods to improve life. Development is thus focused on material production.

The indigenous worldview, of nature as fundamentally alive and sacred, often represented it by the symbol of a circle: the unbroken sacred hoop of life. In many indigenous cultures the basic laws of nature were formulated in accordance with what we now call sustainability: laws of balance, harmony, mutual sustenance, of returning in equal measure for whatever you take.

Understanding the world as a single, interconnected and interdependent living system, the Hopi and Kogi knew that the consequences of the White or Younger Brother's destructive ways would necessarily be disastrous. He took from nature, often leaving scars upon it, produced things and threw away wastes. He did not notice the circularity of nature: that the wastes actually closed a loop, becoming part of his environment, poisoning it if the wastes were poisonous. In the 'sacred hoop' view, there was no concept of waste and whatever was put back into the environment was useful to other species—an excellent life insurance policy for any species, as we pointed out earlier—one followed by the species of mature ecosystems.

No wonder indigenous people noticed the White Brother's failure to restore what he destroyed, and were able to predict the consequences thereof. He mowed down great forests, plowed up the earth to grow food, made gaping holes in it to mine minerals, and dumped wastes onto land and into clean rivers. The Kogi, in particular, could see the mining and cutting of forests below the mountain on which they lived. The more devastation below, the fewer the clouds which used to rise from the forests bringing rain to their lands, which literally dried out before their eyes, forcing them ever higher and closer to the water's source in the dwindling snows.

Indigenous peoples were humble about their place in nature, while industrial society was founded on the conviction that European man was master of all nature and would bring about a Golden Age by conquering, subduing and transforming nature to his own ends. Its founding philosopher John Locke clearly stated, "The negation of nature is the road to happiness," and indigenous people were negated along with the rest of nature.

Only now, when we are in critical danger do we look back to understand the history of the White/Younger Brother's destruction of indigenous cultures, as well as of whole ecosystems, to build his technological world—a world in which nature has been seen only as a supply base and a dumping ground, a polluted world which testifies to the White Brother's failure to respect the Red Brother's sacred Earth wisdom. A world we now recognize as unsustainable.

O   O   O

The image of indigenous peoples as backward and ignorant stands in the way. Their philosophies are still largely ignored; their lands are still under seige as dumping grounds for toxic wastes, in both Canada and the U.S., or for mining. Elders who deserved to be treated as national treasures die in poverty. As of this writing, almost all traditional Hopi elders have died; Roberta Blackgoat, a Dineh (Navajo) grandmother in her advanced eighties, good friend of the Hopi, suffers the theft and deliberate injury of her animals decade after decade in efforts to force her from her home. When the Rockefeller Family reevaluated the basis of their philanthropy some years ago, the president of its well-known foundation repeatedly cited indigenous philosophy for its guiding principles to a better world. Yet indigenous people remain among the poorest in the U.S. and still suffer evictions from their own lands in the name of profits.

Unfortunately, indigenous histories are generally known not through their peoples' own telling, but by anthropological reports. It has been widely assumed that non-technological peoples, many of whom have no written language, do not know their own histories and were not smart enough to develop technologies.

A case in point is that even the urbanized Mayans, Aztecs and Incas with their sophisticated cultures and high arts were seen as backward on the grounds that they did not even invent the wheel. The Incas at least did understand the possibilities of wheels, using them on children's toys, though never for transport. Perhaps burdened slaves were seen as more appropriate to the task of transport. Perhaps the sacred hoop of life was forbidden as a mundane technology. It is instructive to recall that ancient Greeks, even when inventing technology under duress, as in the case of Archimedes' war machines, did not write down the plans. Technology, based as it is on geometry, was considered to be God's sacred art and was forbidden to man, though the Greeks obviously exempted the wheel. The Incas apparently did not.

It is difficult for people born into technological culture to imagine anyone preferring a simple, non-technologically developed lifestyle in a natural setting, with few possessions. Yet, most indigenous people of the Stone Age, as Marshall Sahlins points out, worked very few hours to make a good living. To prefer the leisure time granted by choosing not to be a consumption-oriented society is seen by our own consumer society as laziness; to do without material wealth is seen as deprivation.

Sarah James, a Gwich'in Indian from the northernmost inhabited village of Alaska made the arduous trip to Rio de Janeiro for the Earth Summit of 1992. She described her caribou culture before contact with the white man as rich—rich with family, warm homes and clothing, plentiful food, much time for ceremony, music, dance and story telling, much reason for celebration.

When the white man came to them, he saw only people living in forty degrees below zero weather, with nothing but caribou to provide food, clothing, implements and skin huts. He called them savages and brought them canned goods, junk food, alcohol and real poverty. Sarah beat her caribou skin drum, sang her welcoming skin hut dance song, and smiled broadly as she shouted, "Let's keep Alaska savage!"

Her traditional lifestyle had been truly rewarding—its natural simplicity an integral part of a spiritually rich culture. Her people honored the caribou as their brothers and kept the herds healthy in turn for their gifts. Like hers, most remaining indigenous communities have lost their old values and communal lifestyles, the allure of modern culture pulling strongly, especially to the young. The conflicts within indigenous communities over this issue are heated as efforts to revive traditional lifestyles compete with the trend to assimilation and modernization. One can only hope the traditional values will be incorporated into whatever lifestyles result.

O   O   O

One popular belief we hold about native peoples is that they all had short lifespans due to their backward existence. Indigenous people's lifespans 'B.C.'—a native term meaning Before Contact— as reported by these cultures, were ignored. Instead, statistics on life expectancy were taken after respiratory and other diseases brought in by colonists decimated infants and children, and often older people. The Spanish settlers after contact, for example, gave the average life span of Tewa Indians in the U.S. Southwest as 40. Along with this Spanish missionary statistic, it was reported that half the children died of respiratory diseases before the age of four. That leaves the *average* life span of the survivors of imported disease as 78!

The Kogi bury people who have not reached the age of 96 with strings coming out of their graves so the spirits can leave when their allotted lives are complete. Hopi elders are expected to reach one hundred years and more, and they still do. Shuar Indians say 120 is a normal death age. People from northern white cultures now travel to the Amazon to learn the secrets of longevity.

Unfortunately they are going late, as indigenous cultures are disappearing faster than ever. Many indigenous groups today are fighting or acquiescing to their own extinction. At this writing the Guarani Kaiowa of Jaguapire, in Brazil's Mato Grosso del Sul threaten collective suicide as other Guarani have already done in the face of forced eviction from legally held tribal lands and the murder of leaders. In Ecuador multinational oil companies scramble to take over other tribal lands, of the Huarani, Shuar, and others, extracting oil messily in a country without environmental safeguards. The last pristine headwaters of the Amazon are in these territories, being covered one by one in oil slicks as the natives die from disease and destruction. North American volunteers, shocked by the situations of indigenous people in South America, have tried to help them. Some, like indigenous people opposing these processes themselves, have been murdered for their activities.

Linguists estimate that half the languages of the world are already extinct, and that in one more generation half of those left will be gone. Human diversity is crashing even faster than bioregions are destroyed. The Hopi, the Kogi, the last free-living Aborigines all tell us that they can no longer keep the world in balance through their prayers and ceremonies. The White/Younger brother is too powerful and must now come to his senses or complete the destruction.

In North America, as in other parts of the world, the indigenous survivors of colonial policies forced onto reservations were deprived of their natural economic bases. In Canada, some Indians could get title to their lands only if they 'improved' them by stripping them of trees. In the United States, bulldozers ripped out the pinion trees that provided the sustenance of the Shoshone and the animals of Dine'h (Navajo) shepherds are destroyed even today in efforts at forced relocation of people in order to mine their lands.

Native peoples' religious practices were outlawed until 1978 in a country founded on religious freedom; their traditional governments dismantled, outlawed and replaced by Tribal Councils designed by the U.S. government. Consequently, many native nations are divided by conflicts between such councils and traditional, if 'illegitimate' leadership.

1992, the Quincentennial Celebration year of Columbus' first voyage to the Americas and the year of the Rio Earth Summit with its worldwide meeting of indigenous peoples—in addition to the world's governments and non-governmental organizations—brought indigenous issues into the public eye as never before.

The systematic destruction of native people and cultures is at least now well documented, though still not widely known. Precisely because it is not common knowledge, confusion still exists about what real indigenous cultures were. It is as inappropriate to judge indigenous cultures by the worst behavior we find among their abused and impoverished survivors as it is to glamorize them, to sell their ceremonies, their portraits and their art for profit, with few exceptions giving little or no return to their creators. The point is not to romanticize indigenous people, who have been and are as human as all others, but to acknowledge and learn from their traditional best—from their deeply spiritual respect for and scientific knowledge of nature.

O   O   O

Science is defined by Merriam Webster's Collegiate Dictionary (10th edition, 1993) as "the state of knowing" or "a department of systematized knowledge as an object of study." This definition certainly includes indigenous knowledge. The American Heritage Unabridged Dictionary of the English Language (3rd edition, 1992) defines science as "the observation, identification, description, experimental investigation and theoretical explanation of phenomena." This is a bit more precise, yet a good description of what indigenous people do that is appropriately dignified with the label 'science.'

As defined by the Oxford English Dictionary, science is "the state of knowing", or "knowledge as opposed to belief or opinion," knowledge, that is, "acquired by study." The OED continues explaining that science is "in a more restricted sense: a branch of study which is concerned either with a connected body of demonstrated truths or with observed facts systematically classified and more or less colligated by being brought under general laws, and which include trustworthy methods for the discovery of new truth within its own domain." Detailed as this definition is, there is nothing in it to exclude indigenous science.

While native scientists do not do science in laboratories, they do systematically acquire scientific knowledge through observation, experiment and theoretical explanation in a framework of natural law. Dr. Greg Cajete, a Tewa Indian educator and author from the Santa Clara Pueblo observes that the white man does science in a "low-context environment," isolating phenomena to study them outside their natural context, in a laboratory, while the red man does science in a "high-context environment," studying phenomena within their natural context.

He explains that the reason for this difference has to do with the purpose of science in the two cultures. While both do science in pursuit of knowledge based on real observation and experiment, the white man removes phenomena from their natural context to study them in laboratories because he seeks knowledge enabling him to control nature for his own purposes, while the red man leaves what he studies in place because he seeks knowledge that will permit him to integrate himself harmoniously into nature. Indigenous scientists have always known the participatory universe, while the industrial culture's scientists only recently discovered it, now understanding that the pure objectivity considered so fundamental to doing good science was illusory.

Indigenous science is thus participatory—fostering dialogue between humans and the rest of nature. It is taught to all people, not as something learned in limited years of schooling, but as a lifelong task, though its specialists—such as medicine people who are both researchers and practitioners—spend many years in formal and rigorous training.

Industrial era science consciously and carefully divorced itself from religion for reasons of historical conflict with the church. Efforts to resolve this conflict are being made by scientific theologians such as Thomas Berry and Matthew Fox, who integrate the modern scientific story of cosmology and planetary evolution into religion, by physicists such as Brian Swimme and Stephen Hawking, who comfortably talk about God, by philosophers such as Ken Wilber.

It is noteworthy that on both sides—religion and science—these efforts are seen as *integrating* the separate concepts of God and Nature. In most indigenous belief systems, God as Creator and Nature were never separated, as Creator was the very essence of Nature, just as we discover today. Many indigenous people are puzzled by industrial culture's separative tendencies, with their arguments, for instance, about whether God exists outside or inside the natural world. They ponder our strange separations—our divisions of our lifestyles into separate categories of science, religion, economics, politics, ecology, arts, etc.

One Meshika grandmother, Xilonem Garcia, has said "Anyone who knows how to run a household knows how to run a world." She understands fractal biology—that patterns repeat at different levels, and that a healthy living system must run by the same principles no matter what size and scope it has. Consider the definitions given earlier of ecology and economy as 'organization of the household' and 'operating rules of the household.' This native elder understands that they cannot be separated.

The point of this discussion is not to show one science or cultural pattern superior to another, but to recognize that there can no more be one true science than one true religion. In Chapter 12 we discussed the impossibility of any single true worldview. Science itself is a mapping activity—its theories are testable maps to the underlying reality filtered into our minds through our limited senses. We make many different kinds of actual maps, all valid. We do not expect a pilot to fly by a road map, a driver to drive by a weather map, or a weather forecaster to predict weather from a topographical map. We make our maps for different purposes, just as indigenous and industrial scientists make their scientific descriptions of the world for different purposes.

What matters is which sciences we consider when we want to achieve these varied purposes. Indigenous science may offer little to the design of a radio telescope or new computer operating system, but it may be extremely useful if we want to know how to survive as a healthy part of nature. In Chapter 20 we will see that sustainable agriculture, for example, may better be based on indigenous and traditional techniques than on costly and destructive hi-tech farming.

O   O   O

The political science of the Haudenosaunee contributed much to the democratic Constitution of the United States, as the September 1987 National Geographic Magazine and the later work of work of scholars such as Oren Lyons, Vine Deloria, Jerry Mander and Jack Weatherford, have documented. The founders of the U.S. were refugees from European tyrannies. It was among the Indians of the Haudenosaunee League, whom they called Iroquois, that they—especially Benjamin Franklin—found democratic principles and practices at work. The Haudenosaunee League was a peaceful and democratic federation of tribes that had been historically at odds with each other. Unfortunately, while adopting these Indians' democratic forms—there were no democratic government role models in Europe—the founding fathers left out the equal role of women in governance, as well as the role of children and the sacred contract with nature.

Among the Haudenosaunee, chiefs were selected by grandmothers, who had watched them grow up and knew who would serve their people well. It was also the power of the grandmothers to remove chiefs from their positions if they did not govern and keep ceremony as they should. More generally, women participated equally in all decisions. The Haudenosaunee, like indigenous peoples everywhere, used the sacredness of nature as their guidepost to what should or should not be done by humans. To be sacred is to be inviolable, to be treated with utmost respect. To have a sacred contract with nature is to care for it, protect it, give back for what you take.

Indigenous and traditional communities were necessarily aware of the ecosystems in which they lived. Current anthropological/archeological interest in demonstrating the demise of cultures such as the Maya through their own environmental devastation is inconsistent with their sophisticated level of agricultural knowledge. Droughts not controllable by humans are another matter, and a drought phase lasting hundreds of years is far more likely to have caused the demise of the Mayans.

Some native techniques, such as burning small forest areas, farming them a few years, then moving on, were considered destructive until recently, when our own scientists recognized the value of controlled burning to forest health. The Amazon Kayapo and many other rainforest peoples in other parts of the world carefully included in their gardens plants and trees that would insure the rapid regrowth of forest on each such plot.

Hunting buffalo by driving herds over cliffs is another example used to demonstrate ecological malpractice among Indians. This was indeed a real, though difficult and dangerous practice among a few tribes. Vastly fewer buffalo were killed this way than by the colonists' practice of shooting them by the thousands from trains. All dead animals were fully used by Indians rather than being left to rot as in the latter case. It is true that some young Indians today hunt recklessly when allowed. They fear their hunting privileges may be taken away again at any time. As Haida elder and former leader of the Haida Nation Lavina White (Tthow-Gwelth) said in a speech at a 1990 University of Calgary conference:

To my people, all creation is sacred and our religion is to live in harmony with nature. But... we've had no control over anything, not even our own lives, for a long time. We've been held captive in a reserve system that has no economic base, and we have been unable to live as we should be able to. Before contact—before interference from white men—there was order, and we assumed such order existed throughout the world. ... Some people are worried about how we are going to treat wildlife if we ever get control of our lives again, worried that there would be a lot of abuse. That's untrue historically and...I would like to assure you that our philosophies wouldn't permit us to carry on in that way. Even if there were some people wanting to do that, they wouldn't be able to for very long. If we got our systems back, we could deal with those sorts of things.

O   O   O

Manuel Cordoba, a Brazilian rubber tapper kidnapped as a boy early this century, learned the medicine he practiced all his life from Amazon Indians, as documented by Bruce Lamb. In Cordoba's advanced age, he was called in when doctors failed to cure the chairman of the Medical School at the University of Lima of a terminal illness. Cordoba succeeded, using only indigenous knowledge and medicines. He was offered a professorship at the university, but declined. Today drug companies buy up rights to exploit the Amazon for medicines, patenting them on discovery.

'Invisible' technology appears magical to those not trained to use it, especially in the realm of healing. Many Amazon medicine men use the hallucinogen known as *ayahuasca*, made from several varieties of *Banisteriopsis* vine, to diagnose in detail the physiological problems of their patients, as Cordoba was taught to do. In the hands of trained practitioners, it can be used to unite minds and bodies such that detailed knowledge can be transferred directly among people and other species. Cordoba was even able to telepathically transmit specific physiological information to his wife, who had not taken the ayahuasca. Fred Alan Wolf and other western scientists have now researched these abilities.

Amazon hunters also use ayahuasca visions to locate and communicate with animals prior to hunting them. In Canada, some Indian hunters dream the hunt before going on it, without the aid of hallucinogens. Such a hunter sees the animal willing to sacrifice its life and makes a sacred contract with it in the dream, to give its life. He tells the other hunters how and where to find it, making the actual hunt efficient. He teaches tracking and the making of weapons as well as how to dream. In industrial cultures our great men sometimes say they got information through dreams, as with the scientist Kekule's carbon ring, or some of Einstein's theories. But we write these off as quirks of genius, and show interest only in the formulas that can be written. We do not ask them to teach students to dream.

Polynesian navigators of the Pacific Ocean have traversed its waters for thousands of years without benefit of compass. These navigators knew astronomy for navigation by stars, had sophisticated knowledge of both deep and surface currents, cloud and weather patterns and fish and bird migrations to guide their swift, elegant outrigger canoes over vast stretches of ocean. They were also trained to detect magnetic fields directly in their bodies to give them the compass directions migrating animals have. They could sense their proximity to land, or, as one such navigator said, "stand tall in your canoe, until you see where the land is." Nowadays that is called remote viewing and has been much researched by the U.S. military. Thirty indigenous Pacific nations have recreated their traditional sea-going vessels in recent years in order to retrace ancient voyages using the same techniques.

Much indigenous science is based on centuries and even millennia of observation passed on through time, generating laws of relationship. Hopi observations in geology and meteorology, for example, led to the understanding that underground copper deposits in the Southwest act as lightning rods, drawing down lightning and bringing life-giving rains to the desert. They know that mining can change weather patterns as surely as the Kogi know that deforestation and mining are drying the climate around them so their mountains no longer have adequate snow to feed the rivers on which their crops and lives depend. Both cultures have observed the destruction while the white man saw only the copper and the gold that would bring him wealth.

Australian aboriginal tribes have also observed changes in their outback desert habitat over time—an ongoing trend of decreased rain, increased heat, the reduction of plant and animal reproduction. Of all indigenous peoples, these Aborigines may play the most conscious roles of all in the co-creation of their environments. Many people have by now seen their beautiful dot paintings, but few have understood their significance, which can be learned only by deep relationship with these cultures.

Like Tibetan or Dineh (Navajo) sand paintings, the Aborigines' paintings are traditionally made not on canvases, but on the ground, to be blown away or dissolved back into their place. Each painter is responsible all his or her life to a particular species or element of the land, painting it, in its context, again and again. The painting process itself is a ceremony that takes weeks, and includes ritual song and dance. Its purpose in an Aborigine community is to consciously connect it with all the elements of a particular ecosystem, from sand and rock to microbes, plants and animals, *and help bring it into being from source consciousness*—sometimes called Dream Time—moment by moment.

Such understanding of creation has been completely foreign and apparently superstitious to Euro-Americans. Only now, with the latest discoveries of physics, does it begin to make sense and demonstrate the advanced knowledge we were incapable of comprehending. Deep dialogue between the remaining traditional

Aborigine elders and western theoretical physicists and biologists could lead to new breakthroughs.

This book takes the optimistic position that it is not too late to learn from the ways of nature and the scientific knowledge of indigenous peoples, with their deep ecological wisdom. Cooperation between indigenous and industrial society, based on mutual respect, can help us identify destructive technologies and make useful technologies ecologically sound. It can even lead to advances in our knowledge of how the universe and out own planet create themselves and function, as we have just seen. The White Brother's inventive genius is enormous and capable of solving the greatest problems we face, if it is augmented by the Red Brother's deep knowledge and wisdom.

20

Sustainable Society

Sustainability is now widely discussed, at conferences, in the media, among people in the street. Despite heated debates, many people do not have a clear idea of what it means. This is not surprising, since visions of what sustainability might look like are virtually absent from these discussions except among people such as Bioregionalists and Futurists who are not yet widely represented in the population. On the whole, the debates are based on fragmented worldviews that make it difficult to understand the issues holistically.

People do recognize that the discussion of sustainability has to do with changing the way things are and that it is linked to concern for the environment. Many people are afraid it means ecology at the expense of economy—pitting the survival of endangered species, for example, against jobs and development, as is sometimes phrased "jobs versus spotted owls." It is natural in our culture to think in this way—that for something or someone to gain, something or someone else must lose. This is because we are accustomed to living in what Hazel Henderson has for decades called a worldwide 'win-lose' economy—the kind of economy we discussed in Chapter 16, where we observed how such an economy would kill a living system. Sustainability, in its essence, is about the necessary shift to a win/win economy that would benefit all humanity as well as the other species on which human life depends.

We have seen that the words ecology and economy are related as design and management of a 'household,' and how, in trying to understand the various aspects of human society as parts of a social mechanism, we lost sight of the fact that one cannot separate how our human household is run from how it is organized.

Let's look once more to our physical bodies, seeing their ecology as their organization into interrelated systems—skeleto-muscular, circulatory, digestive, brain/nervous, perceptual, and so on. By what principles do they manage their economy of food intake, of cellular maintenance, of endocrine, plasma, etc. production, of materials and product distribution, of recycling and elimination of wastes?

We know the nervous system acts as a service government—a central guidance system collecting information, monitoring the state of affairs everywhere in the body, working closely with endocrine and blood systems to make sure supplies are appropriately distributed, coordinating the tensegrity movements of bone and muscle, the perceptions from eyes, ears, nose, mouth and skin, regulating body temperature and emotions. Its jobs are far too numerous and complex to mention or track them all. What we can say is that as long as the body is healthy, there is no conflict between its ecology and its economy. It coordinates a win/win economy/ecology in which all parts contribute what they have to offer and all parts benefit equally from the collective economy. No part of a healthy body gains its health at the expense of other parts; there are no such things as rich and poor organs.

If we accept the notion of the living Earth, and the body of humanity as an integral part of it, then we have no choice but to implement a healthy win/win world that can continue indefinitely, which means a *sustainable* world. As long as you are healthy and avoid accidents, you are sustainable for a natural lifetime. In the same sense, a healthy world is a sustainable world. This whole book has been about sustainability, yet the current debate on the subject warrants some further discussion.

In *Earth in the Balance*, Al Gore called for "an environmentally responsible pattern of life." He expressed optimism about the fact that most people now see themselves as part of a global civilization, and that most of the world has chosen democracy as the preferred form of political organization and modified free markets as the preferred form of economic organization. Yet he recognizes that the single most difficult relationship is the one between wealthy and poor nations, clearly stating that the wealthy nations will have to write off impoverished nations' debts and assist their sustainable development to make them partners in a balanced, healthy global economy.

This is a clear case of recognition that sustainability implies a balanced economy of equal partners, rather than an economy in which some nations or corporations gain at the expense of others. Gore recognizes that

Any such effort will also require the wealthy nations to make a transition themselves that will be in some ways more wrenching than that of the Third World, simply because powerful established patterns will be disrupted...the developed nations must be willing to lead by example; otherwise, the Third World is not likely to consider making the required changes—even in return for substantial assistance.

Gore thus cuts right to the core of our global crisis. Our win/lose world is a top-heavy world in which seven percent of the people own sixty percent of the land and use eighty percent of the available energy. Its dominant economy has been and still is based on growth that simply cannot continue its path of destroying ecosystems and creating ever-expanding masses of impoverished and desperate people.

Resource use, population growth, the gap between rich and poor are all proceeding along exponential curves heading quickly toward infinity—and none of them, of course, can reach it. There is no way to have an infinitely large population, to use an infinite amount of resources, etc. So we *know* things will change. Something will alter the direction of change. The question is only, what will it be? Extinction? Other disasters, such as economic crashes or massive technological breakdowns? Or a sudden awakening and resolve to implement sustainability?

One problem with appealing to national governments to shift toward a more equitable world economy is that multinational corporations are now often richer than many nations and have the power to control them. Paul Hawken points out that lobbying for corporate interests in Washington DC is a multi-billion dollar industry with which no other interests can compete fairly. He reports that during a single legislative session U.S. congressmen take 3,000 corporate financed holidays, illustrating how the U.S. government, not to mention campaign financing, is kept in service to business rather than the other way around.

Chief Oren Lyons of the Onondaga Nation uses the metaphor of CEOs of businesses and banks as jockeys on multi-national corporation and bank 'horses,' beating them on to a finish line now visible as a stone wall they will run into, yet not even turning around when one of their fellows falls. What this metaphor portrays is that corporate businesses and banks, including international development banks that serve corporate interests, are in a competitive race of unsustainable world economy. Lyons' stone wall finish line is consistent with our analysis of the death of a living system pursuing win/lose economics. To illustrate our state of denial about our non-sustainable world, Lyons has also used the metaphor of the Titanic, which its owners, crew and passengers all regarded as such a marvel of modern technology it could not possibly go down. Notably, the film *Titanic* became a great cultural hit.

In this light it is interesting to consider historian Arnold Toynbee's observation, after studying twenty-one collapsed civilizations, that what they had in common was inflexibility under stress and the concentration of wealth into few hands. We cannot deny the current stress. Will we remain inflexible in maintaining a system that concentrates wealth to the increasing detriment of most humans?

O   O   O

While human communities were small and in close touch with their ecosystems or bioregions—as we began to see in the last chapter—many of them were able to function in good ecological and social health, often quite democratically, sometimes for many centuries. They functioned much like bodies, with divisions of labor and all parts contributing to each other's welfare.

Helena Norberg Hodge has shown how the rural towns of Himalayan Ladakh, sometimes called 'Little Tibet'—with their three-story white-painted houses, beautiful monasteries, irrigated wheat fields and gardens, herds of animals, festivals displaying their music, theater arts, brocades and silver, crops adequate to support their people in good health, with no poverty—functioned sustainably for many centuries. Buddhists and Muslims lived peacefully together in these communities, with their deep spirituality and strong values.

Despite the considerable property described, the barter economies of these communities were discovered to count as nothing when Ladakh was introduced to the concept of GNP. Only when these barter economies were undermined by the influx of the modern commercial world into this tiny remote country and men were persuaded to leave these communities in order to work for a pittance in cities did the GNP go up. Unfortunately, the result was a great deterioration in the lives of these Ladakhis.

How were the men persuaded into leaving the spiritual beauty, communal harmony and physical bounty of their villages for polluted, congested urban living? As in other parts of the world, roads were built, people were encouraged to stop producing their own food and goods through the import of subsidized grain and other cheap imported goods and the opportunity to earn enough to pay for at least the basic ones.

Motorcycles, TV and videos filled with guns, girls and images of no-work affluence in the West came in to seduce the young men, eroding the economy and the values of rural life. People were told they were backward; that modernization would bring great benefits. Because of the initially subsidized grain, fields were abandoned; school children were systematically taught the values of a market economy, the importance of industrial development.

The living systems of the old communities were thus fragmented beyond repair. Hodge, who lived in Ladakh throughout this modernization process, documents how the happiness of the people plummeted and conflicts erupted as they had to live with difficulty on dreams of a better life that did not materialize for them. As Hodge points out, this single decade of change from peaceful, healthy self-sufficiency to conflict-ridden, miserable dependency reflects in a nutshell the colonial process all over the world, everywhere counted as economic improvement.

The colonial process has been and is always essentially the same—the mining or monoculture farming of indigenous lands by outside owners using local labor for a market economy destroys self-sufficient, independent and secure community, whether the lands are seized outright or bought cheaply. Removing the men to provide a work force leaves women and children with more work and a ruptured society ever more dependent on cash. The old community rules and values cannot be sustained; population is no longer controllable. Later, this destruction is compounded by urbanization and industrialization, which makes people totally dependent for their livelihood on paid work and impersonal institutions, such as credit banks, supermarkets, hospitals, etc.

In her interviews with African village women, population expert Perdita Huston found that grandmothers were socially and economically better off in intact tribal societies than were their granddaughters in modern economies. In Ladakh, the downward change has happened in a single generation. In Africa, the process has been going on longer. But in any case, the majority of people of these formerly healthy living systems end up poor or destitute on barren land or in urban slums. They have become part of a world economy in which they serve as cheap labor and market outlets if they are lucky. Increasingly, they are left out of even these slim benefits, desperately poor in huge urban slums, on the edge of starvation throughout their lives, many never reaching adulthood.

According to world futurist Rashmi Mayur and TV documentaries, many millions of children in Bangladesh and India under the age of ten are enslaved up to nineteen hours a day, seven days a week in factories making goods for export to the U.S. Making consumers conscious of these conditions has created a groundswell of protest—exactly what is needed to change them.

A historic1994 *Atlantic Monthly* cover story by Robert Kaplan—illustrated by a world on fire—documented the devastating reality of desperate poverty imposed on peoples in Africa, Asia and South America. Kaplan points out that to believe things are still well in the world one must ignore three-fourths of it. If we see the situation realistically, we know it is entirely unsustainable, causing enormous and unnecessary human misery.

We are, in fact, in the same desperate situation bacterial colonialism led to a few billion years ago. Yet the nucleated cells they devised as sustainable solutions have survived and flourished some two billion years in a myriad evolved forms. They are so sustainable that no other kind of cell was ever needed to replace or improve on them. The same cooperative communal solutions they found are open to us.

O   O   O

It is clear, then, that money is driving out world—that money, not the good life Bacon foresaw, has become the whole rationale for our economies. We measure the health of our economies only by the amount of money flowing in them—the GNP. Can we not see there is something wrong with this measure? Terrible oil spills increase the GNP due to the money flowing in to clean them up; increasing expenditures on other human and environmental remediation do the same.

Hazel Henderson has asked us for decades why money should be the measure of our social health, while we ignore the real costs of destroying nature and lives, as well as the real assets of all the creative unpaid labor volunteered to raise and maintain families and communities around the world. Her Quality of Life Indicator scales were pioneering efforts many have followed, though they remain to be seriously applied to national and corporate economies, as they must be to measure our progress toward a sustainable world.

How did the concentration of wealth become so dominant a force in what we call democracies? Since wealth is generally defined in terms of money, let us look further at money. Belgian banker Bernard Lietaer has pointed out that money is simply an agreement on the value of some medium of exchange used to facilitate relations among the producers and consumers of an economy. In an equitable democratic society, representative government would issue or withdraw money from the economy only to balance these relations. The interests of the entire citizenry in determining how to balance the economy would guide such a government. This was the general idea held by the founding fathers of the United States, who warned against implementing a debt-money system, known to be detrimental to all but the lenders since ancient times. For this reason the U.S. Constitution was written to make Congress the only body that could coin money.

Jacques Jaikaran raises the interesting question of why the United States Congress gave its constitutional right to issue money away to a private banking system with the public-sounding name Federal Reserve Bank at its core, forcing the government itself to borrow money at interest. He describes how the debt-money system implemented by these banks functions to funnel money and property from the poor to the rich, thus fostering the process of a win/lose world that is fundamentally unsustainable.

Money is concentrating with unprecedented speed in the hands of a small world elite, as it does in the hands of one player of every game of Monopoly. We are all caught in this giant monopoly game, which cannot go on much longer, by reason of impossible exponential curves. Something, we can be sure, will soon break or shift dramatically.

When it does, will the people of the world effectively demand a different and truly democratic economy that does not destroy the living systems of nature and people within nature? There is at present a trend toward equity money in place of debt money, to keep the system going longer. Equity money means more people in upper classes will share ownership of businesses including banks, but this does not solve the problem of vast numbers of poor people who will be as disenfranchised as ever, if not more so.

More promising is a big groundswell of alternative currencies around the world, from the computer tracked Local Economic Trading Systems (LETS) pioneered in Canada by Michael Linton, the Mexican Tlaloc and U.S. Ithaca Hours trading notes now copied by many U.S. communities, to airline frequent flyer miles and volunteer community services hours in U.S. states as well as in Japan's elder care trading system. Lietaer calls these the growing Yin economy that is coming to balance our monetary Yang economy. It is instructive to note that local communities across the United States survived the Great Depression beginning with the stock market crash of 1929 with exactly these kinds of local barter currencies—later stopped as 'inefficient,' though legal. Now the world's people—in the U.S., Australia, Mexico, Europe, Asia and elsewhere are implementing them *before* disaster strikes. Perhaps we *are* becoming more intelligent as a species.

O   O   O

Did modern medicine create the problem of overpopulation by saving so many lives? It is true that colonizers brought diseases to which natives had no resistance and which thus decimated whole populations. It is also true that modern medicine has worked to combat such diseases, which may compensate for the lives lost before the medicines existed, but it is not an adequate explanation of overpopulation. Dramatic increases in food supplies when people are used as labor to produce food in quantity for markets has also been cited as cause for overpopulation, yet the very countries producing food for export, as we have also seen, are those with the highest starvation levels. Wherever people die at high rates is where we find them replaced in ever greater numbers.

Certainly urbanization, sanitation, technology and agricultural monoculture did increase the world's human population. Before these mixed blessings of colonialism, overpopulation was rare. Traditional communities with subsistence lifestyles consciously regulated their population size. Indigenous and traditional peoples survived for many thousands of years without overpopulating because the people of these societies knew their bioregions, understood well how many people their land could support. If populations grew greater or less than optimal, they adjusted social practices, such as how long nursing mothers were off limits to men, how many husbands or wives could be had, how many people remained celibate in spiritual life. Most such societies also had knowledge of herbal birth control and in some cases selected infanticide was practiced.

Overpopulation, like other social problems, occurs when communities in sound ecological balance with their surrounding world are destroyed and that balance is lost.

Population discussions *must* address the problem of resource overconsumption in rich countries, rather than focusing only on numbers of people. In fact, the reason we worry at all about population size is because the consumption of resources has become unsustainable. This raises the question of whether a bigger population problem arises when many people live on few resources or when few people live on many resources.

The IPAT formula (Impact = Population x Affluence x Technology) of Paul and Ann Ehrlich, showed us that the resource drain of Americans is far greater than that of Bangladeshis. The average American uses somewhere between forty and seventy times the resources of a person in a poor country such as Bangladesh. Multiplying the U.S. population by the conservative figure of forty means the U.S. population pressure is that of over ten billion poor people! This is very important to consider when deciding who is 'overpopulating' the Earth. It is abundantly clear that all the world's people can never live as Americans do today. Thus the reduction of consumption is more pressing than the reduction of numbers, making the Voluntary Simplicity movement we mentioned in Chapter 18 very important.

Another major sustainability issue is pollution. Looking back again at Gaian evolution, we recall that humans are not the first creatures to threaten their own and others' extinction by way of resource depletion and pollution. In considerable detail, we followed the ancient bacteria as they survived similar crises by reorganizing themselves and their living systems repeatedly.

We also saw that species living now can exist only because the Earth spent billions of years burying atmospheric carbon in forests and underground. We noted that cutting and burning these forests and fossil fuels reverses the planet's system for keeping atmospheric conditions and climate conducive to species health. It is not a sustainable way to live. It is the way of an immature species that gobbles up all available resources, like the weeds that take over land along our highways or in abandoned fields, where we have destroyed mature ecosystems.

Technological production is natural to the human species, but must be reevaluated and revised in a goal-setting context of healthy survival. We have discussed the ability of mature ecosystems to clean up considerable human pollution, if they remain healthy. Destruction of forests, seashores, water tables, ozone layer, etc. make it impossible for the Earth to perform that cleanup. But perhaps it would be wise for us to reconsider the whole notion of pollution and cleanup.

Earthlife as described in this book is fundamentally and necessarily based on recycling. Because the need to recycle our human products, lest they choke us out of existence, has become so urgent, a new branch of biological science is finally looking at nature's recyclers. It has now been estimated that sixty percent of all species are 'recyclers.' While this new science at last vindicates the vultures, worms and microbes we have looked down on for so long, it is actually misleading. The natural world is not divided into producers and recyclers; all species are both to varying degrees.

In a mature, balanced ecosystem, there is no waste or pollution, no special cleanup required. The principle of mutual consistency suggests that a healthy species insures its survival by putting out only quality material. Quality material is something useful to others. It is only our industrial culture—immature from an ecological/evolutionary perspective—that creates polluting wastes and must then clean up. But it is becoming increasingly evident that adding more technology to clean up ever increasing wastes is a losing battle and cannot lead us to sustainability.

Paul Hawken urges us to go back to the drawing boards and redesign all our products so that they are either consumable or recyclable. It is not a matter of saving the environment, he says, but of saving business. Hawken proposes that if companies producing non-consumables were only allowed to lease them, and not to sell them, ultimately being responsible for their disposal at great expense, they would quickly redesign them to be recyclable.

With the considerable knowledge of living systems and their dynamic ecological balance we now have available, it is up to us to work *with* life *for* life—eliminating waste as a concept and as a reality. Positive efforts are already reflected the growth of Industrial Ecology as a field. The 3M company is an early pioneer in developing new designs to eliminate waste; some car companies are following suit, especially quickly in Germany. Interface Carpets implemented massive savings through recycling. Green industrial parks where each industry's wastes feed another industry, as in Kalundborg, Sweden, are becoming important models for the rest of the world.

In Chile, a study showed that more energy could be saved through energy efficiency measures than would be produced by six new dams being built on the Bio Bio River, yet the project continues. Meanwhile, a few oil companies, such as Sunoco and BP-Amoco are taking steps toward the inevitable phase-out of the oil economy, investing in solar energy and other alternatives, requesting pollution taxes for the entire industry, etc.

Eliminating waste is more generally about reducing our impact on the planet—giving up the wasteful consumer lifestyle in which we define ourselves by unnecessary accumulations of goods, rather than by human values. It is also about implementing accountability for restorative behavior and using renewable or permanent energy sources to make what we do need, as Amory Lovins of the Rocky Mountain Institute effectively demonstrates can be done. In

Davis, California, energy consumption has already been halved by the sound energy/ecology practices of its citizens over the past few decades. Davis boasts as many bicycles as people, streets now shaded by the thousands of trees they planted themselves, farmers' markets year round, various cooperative housing projects. Most important is the recognition that living with energy efficiency and working with nature provides people with a better life. The furniture in the Davis City Hall had to be rearranged to permit greater citizen participation. When people feel needed and are able to make a difference, they become governors by choice, and they learn to govern well.

The city of Curitiba, Brazil—widely known as the world's first eco-city—has carried the Davis experiment to new heights, as documented by Bill McKibben, who also showed us how much of the Adirondack mountains of the U.S. have been reforested and how the state of Kerala in India has pioneered rural sustainability in a culture now almost one hundred percent literate, with fine health care and education systems while it remains materially very simple.

For all that is not yet being done toward sustainability, we are becoming very aware as a whole world culture that it is the only way to gain a healthy future and we see more and more pioneering efforts in that direction.

O   O   O

The agricultural industry is also beginning to shift its unsustainable practices. Let's look at its story.

Hi-tech monocultures, intended to solve world hunger, have been disastrous in many parts of the world—the World Bank admitting its failure in funding the making of deserts where they had intended gardens. While the affluent world can eat whatever they want from anywhere in the world year round, arable land is being destroyed and eroded by unsustainable practices and ever larger non-affluent populations are ever hungrier. Physicist Vandana Shiva documented the Green Revolution in India, tracing the development of nitrates-dependent agriculture to the need to maintain the production and profits of nitrate explosives factories after the Second World War. Nitrate dependent crops were deliberately bred for this much-touted Green Revolution. The resulting yields of rice per hectare, for example, were shown to be far greater than those using traditional methods— but the measures were misleading because they ignored the fact that the same hectares were not only producing rice traditionally, but fish, pigs, vegetables, fruit, fertilizer and mulch on soil and in water that remained healthy with no chemical input. None of that was counted in the comparison. In fact, Green Revolution fields over wide areas of India became salt deserts, as the World Bank acknowledged.

Hi-tech agriculture was sold to us with other misleading statistics. We were told, as one success story, that a single U.S. farmer at the turn of the century could feed only four people, while with hi-tech agriculture he could feed seventy or eighty or more people. Such statistics ignored the veritable army of people and resources producing the chemical herbicides, pesticides and fertilizers, the rapidly obsolete heavy machinery, the fuels and irrigation systems, the genetically engineered sterile seed that must be bought annually.

In fact, the natural farmer at the turn of the century produced ten calories of food energy for every one calorie of energy input and kept his soil and water table healthy, while the present-day farmer puts at least ten calories of energy into his farm for every one calorie of food he gets out. Meanwhile his land is increasingly impoverished, thus destroying the very basis of his livelihood. Hi-tech agriculture must be counted enormously inefficient and energy wasteful.

It is also argued that hi-tech agriculture is necessary to produce the sheer volume of food required by today's populations. The case of India above belies this, as do the production figures of restored traditional techniques. In the Philippines, one of the countries where hi-tech Green Revolution techniques were pioneered, the restoration of traditional organic rice-growing methods proved superior in quantity of production. Bill Mollison's Permaculture techniques, which can feed many people on very small plots of ground, adapted much indigenous and traditional knowledge, is now taught in over seventy countries and the program cannot train teachers fast enough to meet the demand.

A century ago, a British agricultural expert toured India to see how he could best advise Indian farmers to improve their agricultural practices. His conclusion, reported in *The Ecologist* magazine, was that the Indian farmers had more to offer English farmers in the way of advice, because they knew so much about soil composition and health, pest control, water management, crop breeding, and all other aspects of agriculture. They were highly knowledgeable and productive, failing only when they lacked access to natural resources.

Oswaldo Rivera and Alan Kolata have reported on the restoration of the ancient (400 to 1,000 A.D.) pre-Inca *waru waru* or *chinampa*-type agriculture in the altiplano of Peru and Bolivia. It increased local annual production from the norm of 2.5 tons per hectare to 40 tons in only five years with no chemical fertilizer or pesticides and very little work beyond filling the ditches between soil mounds through sluice gates annually and planting seeds without plowing. In this system nature creates its own fertilizers, the canals becoming a nutrient sump for nitrogen and phosphorus through colonization by fish, birds and water plants. The system's automatic irrigation is also a form of climate- control that prevents freezing in winter. The usual crops were varieties of potatoes, grains such as maize, quinoa and amaranth, legumes, etc. Now winter wheat, barley, oats, turnips and other vegetables have been added, including even lettuce at 2,300 meters altitude.

Few indigenous peoples of the Americas used plows, which are a major cause of soil erosion. All over North and South America, not to mention other parts of the world, indigenous agricultural peoples without the urban social organization of the Inca were equally sophisticated in smaller scale agricultural practices, as Darrell Posey and others have documented and as mentioned in the previous chapter. Each culture understood, through scientific researches over centuries of time, how to breed and grow food, medicine and building material crops appropriate to their biore-gions in sustainable ways.

Poisoning from pesticides and herbicides is normal today in all our bodies. Pathogen contamination (Salmonella, Listeria, E. Coli, etc.) of mass-produced food is a serious problem. Ralph Nader has told us for more than a generation of the pollution in our meat and other food; by now other studies abound. According to the Institute for Science in Society in Washington DC, every year in the U.S. alone, where food supplies are cleaner than many other parts of the world, millions become ill and thousands die from pathogen contamination of meat, eggs and other foods, and this cause, while medical and productive losses therefrom are counted in the billions of dollars. The most popular fast foods, chicken and ground beef, are among the most dangerous food items.

BBC film team John Seymour and Herbert Girardet asked a California tomato farmer why he grew tomatoes for his family in a special kitchen garden when he had thousands of acres of them. He replied that if they understood what agricultural poisons were built into every cell of every tomato grown in his fields—his kitchen garden was organic—they would neither ask that question nor ever eat another canned tomato in their lives! He then explained how he was trapped in this method of production by deep indebtedness for machinery, chemicals, irrigation and the need to meet contract quotas.

Another reason given in support of hi-tech agriculture is the low price of supermarket food in relation to organically grown food. Yet every year at the annual Bioneers Conference in San Francisco, it is demonstrated that organic food can be grown more cheaply than hi-tech food. We are never told the real cost of supermarket food, most of which is government subsidized, but clearly it is far less expensive to grow labor intensive organic food, which could create much-needed employment.

Many urban dwellers today say they would go back to the farm in preference to living in dilapidated, hi-crime inner cities. Holland and Denmark are eliminating chemical agriculture and the latter bonds professional farm sitters so that farmers can spend time in cities. Communications technology can bring urban advantages to country life. And agriculture itself, even if low-tech and natural, would not be as difficult as it used to be if more indigenous (e.g. no-plow, no-till) methods were used.

Vandana Shiva has also documented the 'piracy' of patenting plants and the dangers of genetic engineering of agricultural and medicinal crops. Seeds developed over thousands of years by indigenous peoples or peasants and traded or gifted in sacred ceremony have been defined, for example, as 'primitive cultivar' until brought into a laboratory, genetically altered and then patented for ownership. Under such an agreement the indigenous peoples or peasants who developed the seeds can be fined for planting them unless they buy the seed from its new owners.

Genetic engineering has become the source of great public controversy around the world, as Richard Heinberg and others have documented and discussed. Our experiences with DDT, originally promoted as good for us all, and with nuclear energy that was supposed to solve our problems but brought us Three Mile Island and Chernobyl, has made us leery of yet another panacea.

Instead, in Europe and America, the increase in public demand for healthy organic food has risen dramatically over the past few decades, creating a very significant shift in agricultural production. Before the turn of the millennium, organic food production in the U.S. had become the only agricultural growth industry, and in California, public schools began to implement organic food lunches for children. Meanwhile, in Europe, the public outcry against genetically engineered foods had caused England to make new labeling laws even for restaurants and the European Common Market to reject them entirely as imports, and test fields for such crops were being burned around the world in protest.

Perhaps the engineering of poor cloned Dolly the sheep woke us out of the sheep-like apathy Dostoevsky bemoaned, and we are at last taking responsibility for the great freedom Gaian evolution bestowed upon us.

O   O   O

Bioregionalism, as described in Chapter 16, is one such vision. A bioregionally organized world would most likely include various forms of scientifically integrated permaculture derived from indigenous and traditional agriculture along with appropriate technology for other aspects of life, from communications to housing, medical care, etc. Local production would meet as many needs as possible for food and other goods, with imports determined by democratic discussions. Community would naturally become vital again in such settings, and local culture would flourish, while also exchanged with other regions.

Urban areas would still be desirable and necessary for efficient technological production and other activities and institutions, such as research institutes whose knowledge could then be made available electronically to all. Many people are working on sustainable urban designs that integrate gardens and use clean, efficient energy and public transport. Questions are being asked about optimal city size, about which technologies will be appropriate, about design to eliminate waste. Communications technology would link all bioregions, becoming the central nervous system for the body of humanity. And we *can* make our computers and other equipment without harming either the people making them or the environment.

Bioregionalism is consistent with grassroots democratic movements that are cropping up all over the world, creating new local self-sufficiency systems with their own currencies. From huge housewives cooperatives in Japan to sustainability movements in the hi-tech Silicon Valley world, ordinary people are taking control over their lives into their own hands and practicing local democracy.

Many people wonder how long we have to turn things around. It is really not a question of some critical turning point, but of nurturing more viable systems even as the old ones decay. One metaphor for our changing world is Norie Huddle's story of a caterpillar's metamorphosis into a butterfly. After consuming hundreds of times its own weight daily as it munches its way through its ecosystem, the bloated caterpillar forms its chrysalis. Inside its body, new biological entities called imaginal discs arise, at first destroyed by its immune system. But as they grow more in number and begin to link up, they begin to survive. Eventually the caterpillar's immune system fails, its body goes into meltdown and the imaginal discs become the cells that build the butterfly from the spent materials that had held the blueprint for the butterfly all along. In just this way, a healthy new world, based on the principles of living systems, can emerge through today's chaotic transformation.

There are as many ways to build a new world of living systems as there are creative people who want to do it! Remember that we have seen all evolution as an improvisational dance. Each person, as an imaginal disc, can contribute to the process of today's metamorphosis in some unique way. What matters is that we all understand the Earthdance and the healthy features of living systems at their best. From there we need only the will and the love to create a better future for all living beings.

O   O   O

Before we leave this discussion of sustainability, let's look at a marvelous reinvention of ancient biotechnology that is changing our world faster than anything else we have discussed: the WorldWideWeb, or Internet.

Just as bacteria learned to trade information from one to another worldwide, like frenzied traders on a stock exchange floor, as Lynn Margulis has put it, so we humans have suddenly set up our own worldwide information exchange, qualitatively different from our international telephone calls, though using the same lines. It all began when the army sought a way to get messages through phone lines in case parts of their web was knocked out in warfare. Packet switching was invented, a distributed network of nodes permitting snippets of messages—not unlike bacterial genes—to independently seek open paths from A to B along working routes and then be reassembled at their destination. Victor Grey's *Web Without a Weaver* is a very readable history of the Web, showing clearly why it is different from all previous systems, how it caught on, why all efforts to control it have failed and what it means to our future. *The Cluetrain Manifesto*, by Rick Levine, Christopher Locke, Doc Searls and David Weinberger is an exciting book showing how the Internet is changing our whole society, including our economics, through conversation.

From an evolution biology perspective, what is most exciting about the Internet's nature and explosive growth, is the way its multiple designers and users have unwittingly adopted all the organizational design and operating principles of living systems—that is, their ecology and economy as we have defined them. In Chapter 15, we pointed out that machinery is always an extension of humans. Now we can see that Internet players—who come increasingly from all walks of life and all persuasions—have self-organized, hooking their computers to phone lines through modems, a real and apparently viable living system. It grows in size and complexity daily—generating, processing, sorting and distributing almost unthinkably large amounts of material around the clock and around the globe at ever-greater speed.

For a surprisingly long time, the Web remained a growing conversation by e-mail and chat rooms, backed up with mushrooming websites filled with all manner of information, despite corporate attempts to sell goods on it. Its players were simply too focused on conversations, on hacking at browser designs and developing all

manner of new software to create and send each other artistic messages, music and animations to even think of consuming goods.
Then, quite suddenly, new kinds of businesses began to sprout up
on the Web, with bookstore Amazon.com the first howling success.
Within weeks after its launch, the Web was going crazy with new
venture capital and businesses, soon throwing Wall Street for a loop
with the hottest stocks on the market.

Many of the people who developed Web businesses did so in
desperation when they were fired in massive corporate 'downsiz-
ings.' As Wall Street Columnist Thomas Petzinger has shown, that
tactic backfired, never benefiting the corporations—on the contrary, they suffered a new and stiff competition from their own
ousted employees. And the explanation lies in understanding living
systems, as Petzinger was astute enough to recognize, writing head-
lines in *The Wall Street Journal* on February 26th, 1999 over an
article excerpted from his book on *The New Pioneers*:

A New Model for the Nature of Business: It's Alive!
Forget the Mechanical, Today's Leaders Embrace the Biological

Let's look at this more closely. The corporate world had been
organized in the context of the old mechanical worldview, with top-
down command-and-control hierarchies engineered to keep people
in their departmental boxes, doing only the jobs prescribed.
Management was about keeping them there and keeping them on
their toes lest they be fired. The conversation and creativity people
were finding on the Web was not permitted. Business consultants
went through waves of fads for making business work better—Total
Quality Management (TQM), Total Performance Accounting
(TPA), and so on—but nothing seemed to improve the situation,
and downsizing was a last resort at streamlining.

Those that created businesses on the Web were not seeking a bottom line of profits so much as an enterprise that would succeed because it met human needs—whether for hard-to-get widgets or used books from second hand bookstores. It worked. There were instant computer connections with suppliers and equally fast connections to buyers—it didn't matter where they were, things could be ordered and sent into the mails directly to consumers as fast as the buyer made a decision. The new entrepreneurs went after everything from goods sitting in corporate warehouses, left over last year's models, to unused radio and TV airtime and auctioned them all off. Before long we can expect a flurry of Internet currencies, and a way to interface them, possibly competing effectively with the dollar economy.

What we are seeing is a massive liberation—the discovery of what it is like to function as a creative living system after being kept in the prison of mechanistically engineered schools and workplaces. The corporate world cannot ignore it. In fact it quickly began hiring savvy young people with Web experience to teach them how to talk to people on the Web, how to market to them in personal ways, how to win "captive eyeballs." In other words, the Web has demonstrated the power to lead the behemoth corporate world by the nose into *its* practically homemade, until recently shoestring budget, world.

How can we identify the operating Web principles that made this remarkable feat possible? Let's look at some of the organizational and operational features we can observe in *biological* living systems, be they cells, bodies, communities, ecosystems or our world economy. Observe that the Internet displays *all* of them, while most corporations include very few of them. *This shows clearly that the power of the Web is the power of Life itself!*

Organizational and Operational Features of Healthy Living Systems

COMPLEXITY (diversity of parts)
SELF-CREATION (autopoiesis)
SELF-REFLEXIVITY (autognosis—self-knowledge)
SELF-REGULATION/MAINTENANCE (autonomics)
RESPONSE ABILITY—to internal and external stress or change
EMBEDDEDNESS in larger holons and dependence on them (holarchy)
INPUT/OUTPUT of matter/energy/information from/to other holons
TRANSFORMATION of matter/energy/information
COMMUNICATIONS among all parts
EMPOWERMENT—full employment of all component parts
COORDINATION of parts and functions
BALANCE OF INTERESTS—negotiated self-interest at all levels of holarchy
RECIPROCITY of parts in mutual contribution and assistance
CONSERVATION of what works well
INNOVATION—creative change of what does not work well

A comparison of these principles with those by which corporations operate makes the point more clearly.

MECHANISM (Corporations)	ORGANISM (The Web)
ALLOPOIETIC	AUTOPOIETIC
INVENTOR CREATED	SELF-CREATED
HIERARCHIC STRUCTURE	HOLARCHIC EMBEDDEDNESS
TOP-DOWN COMMAND	HOLARCHIC DIALOG/NEGOTIATION
SYSTEM ENGINEERED	SYSTEM NEGOTIATED
REPAIRED BY ENGINEERS/EXPERTS	REPAIRS ITSELF
EVOLUTION BY EXTERNAL REDESIGN	EVOLUTION BY INTERNAL REDESIGN
EXISTS FOR PRODUCT OR PROFIT	EXISTS FOR HEALTH AND SURVIVAL
SERVES OWNERS' SELF INTEREST	SERVES SELF/SOCIETY/ECOSYSTEM

The influence of the Web on the corporate world is enormous. Will the corporate world follow suit in coming alive, in reorganizing itself from mechanism to organism, as a few of its pioneers have already done? After all, life works, and corporations are made of people who are capable of reorganizing themselves! Why would they want to continue behaving like machines once it is obvious that life works better.

What would a corporation look like when it makes the change? Would it still be a corporation? David Korten, in *The Post-Corporate World* and Jeff Gates in *The Ownership Solution* offer creative ideas on this subject. The new organization's interests will clearly have to be compatible with the interests of its own stakeholders, their families, and all society. Whatever it calls itself, we can guess that it will:

Ø Be autopoietic and holarchic

Ø Create value both internally and externally for all constituencies

Ø Make "shared destiny" moral contracts with employees and society

Ø Shift from absentee shareholders to involved stakeholders

Ø Ensure the recycling of all products not consumed

Ø Treat other organizations as respected equals (friendly competition)

Ø Have triple bottom lines: profits, social development, ecosystem health

In this transition lies our greatest hope for becoming a mature species in time, for corporations are the most powerful human institutions on the planet today. They are the only ones with the resources and ability to make the transition from our acquisitive species adolescence to wise maturity in time to avoid massive disaster.

21

Cosmic Continuation

Life, as we have understood it in this book, is the fundamentally self-organizing, or autopoietic, activity of our planet and our universe—the name of the game, we might say. Universal matter evolves into holarchies of living systems, driven by energetic interactions between great and minute events, between the tendencies of matter/energy to differentiate and reintegrate at ever-new levels of organization.

In this view of life, to recap, matter/energy arranges itself into bounded but interacting living systems on galactic and super-galactic scales, as well as more locally on the scale of our planet, from its entirety to its microscopic bacterial domains. From our present perspective and limited knowledge, it appears to us that planetary life has evolved the most active and complex systems.

In the older, still very active mechanical worldview, life was understood as an incidental and accidental part of the universe, rather than as its essential tendency. In this view, the lifeless universal mechanism had already been grinding along and running down for more than ten billion years since its Big Bang when some of its non-living matter—on at least one rare planet—was accidentally converted into living matter. This was the only way life scientists could fit their discoveries into established physics models of our universe.

Because astronomer-physicists founded modern science, their mechanical mathematical models of the cosmos were accepted as the basis of all science. Physics is still considered the most basic science—the one responsible for explaining non-living matter, the one responsible for explaining how the cosmos is formed. We can scarcely guess how far the organic worldview might have been developed by now if biologists instead of physicists had played the leading role in science—if physicists had had to fit their discoveries into the model of an organic, living universe.

Biologists, who work with living organisms, seeing them reproduce, develop as embryos, care for themselves and make more of themselves, have often found it hard to fit life into the mechanical worldview. But biology simply did not have the status of physics, and so biologists had a hard time challenging the physicists' models of nature. Only now are physicists really coming to the fore with the implications of consciousness in quantum theory, and with experimental evidence to support that implication. But most of biology still lags behind this new physics, in rather Newtonian models of molecular assembly.

Now that we can see Earthlife as part of a self-creating galaxy, as planetary crust transforming itself into a web of creatures and environments, it is clear that living things are *not* built up from pieces as is machinery. Life forms are not assembled by accident from molecules here and there on some non-living planets and then in turn assembled into ecological systems. Rather, some whole planets develop the metabolism of living beings, coming ever more alive in the great flow of energy between their stars and themselves, gradually packaging their crustal material into ever more creatures that weave their own changing environments.

As yet we don't know what role living planets play in their galaxies and within the larger systems of our universe. We are only now discovering the first planets of other star systems, and as yet we know nothing of their life forms. As a matter of fact, we are still discovering new life forms on our own planet.

Our galaxy, we can see, has differentiated itself from a protogalactic cloud into a complex system of nucleus, stars, star systems, clouds, planets, comets, and other parts. Most likely, only a few of many planets—as with the spores, seeds, and eggs of Earth's life forms—come to develop. But a few of the vast number of planets likely to exist in our universe would still be a vast number.

If planets that do spring to life evolve like the smaller living systems of Earth we have observed, then their individuation—as they are born from supernovas—will also lead to tensions, conflicts, resolutions and cooperation as some great cosmic body or being. Scientific models of our own early Earth are still changing. Was it surrounded by clouds of methane and ammonia? Did it rather have an atmosphere in which carbon dioxide predominates, making early organic compounds more difficult to produce locally? Were living molecules imported by comets, asteroids and meteorites? David Deamer has extracted organic meteoric material that forms cell-like membranes, as well as light absorbing pigments that appear to be precursors of chlorophyll. Christopher Chyba thinks the dust that formed our planet—and continues to bombard it daily—could already have harbored such materials.

There are endless mysteries still to solve. A great variety of organic molecules, for instance, come through interstellar space, apparently from the center of our galaxy. Yet we don't know how they can be formed there or even what this center is made of, because great clouds which our telescopes cannot penetrate hide it from us. Only now, as we said, do our telescopes begin to reveal planets belonging to stars other than our Sun.

Our present understanding is as though we were mitochondria in our own cells, trying to understand the organ galaxy and universal body we are part of without being able to see or even guess very much of what these larger holons are all about. Our actual mitochondria may understand us and the cosmos better than we, because a reflective brain looking only outward for information does not limit them. Humans practiced in looking inward, through meditation, have revealed a great deal about the universe, most of which is not accepted in science, because it cannot be subjected to experimental measurement. However, as science progresses, we find more and more confirmation of ancient scientific cosmologies, such as the Vedic.

One of the central elements in Vedic science is that reality, including matter, is created by consciousness, and that matter itself is a created illusion, rather like the matter in our dreams. Now western physicists, as Sally Goerner points out, are also coming to understand matter as an illusion of energy in motion. Physicists have long been talking about fields—traditionally taken to mean all matter, or mass, and energy in a particular region. Einstein's $E = mc^2$—meaning there is a relationship between mass and energy that is mediated by the speed of light squared—was taken to be a conversion formula for matter into energy or vice versa. But more recently some physicists tell us that the interaction of massless electric charges within an electromagnetic field creates the *appearance* of mass. In this scenario, Einstein's formula becomes "*a statement about how much energy is required to give the appearance of a certain amount of mass.*" (*The Sciences*, Nov. 1994, p. 26)

Consider a universe of pure energy with the appearance of material reality. To have an *appearance*, there must be an observer, and as quantum theorists pointed out long ago, in a completely interconnected universe, consciousness anywhere means consciousness everywhere. Now non-locality tells us that anywhere *is* everywhere! In fact, it would seem that energy itself, like matter, is an 'appearance' of consciousness. This certainly fits with out previous observation that no human—scientist or other—has ever had any experience outside of consciousness or outside of the eternally present moment.

Thinking things through in this way we see how limited our worldviews have been. And yet, for daily existence in our reality, our usual concept of matter is still practical. When physicists told us that chairs were made mostly of empty space, they did not begin to collapse beneath us. Learning that they are illusions of consciousness will also not cause them to collapse, since our consciousness creates ourselves from the same 'stuff' as the chairs. Note that we can sit on chairs very well even in our dreams, causing eastern philosophers to speculate on what is waking experience and what is dream.

Nevertheless, there *will* be enormous effects of learning that our consciousness creates our reality—that our assumptions, our beliefs as individuals, as societies and as humanity are the basis of the world we produce for ourselves and co-produce together, along with all living systems, from moment to moment. Jane Roberts has given us the most complete description of how our world works in these terms in her Seth books, more and more corroborated by physics. One of Seth's more challenging questions is, how much we can really learn about the deep nature of the universe by measuring matter with material instruments? If we chase ever smaller material particles with material measurement devices, he says, we create the particles we find from consciousness as we create the rest of reality, and can play the game till we tire of it and learn to study consciousness itself.

Non-Euclidean geometries and the theory of relativity broke through the limits of Euclidean geometry and classical physics. The organic worldview overcomes the limits and lifelessness of the mechanical worldview. A consciousness worldview will give us even greater perspective on our creative universe and our role at its leading edge. It will also give us the freedom and power to recreate our world in ethical integrity, from a place of community and love.

Science does at times reach out boldly into the new, but on the whole it tends to be conservative. It has taken half a century to accept the conscious universe implications of its own quantum theory, and as much time to accept the implications of DNA as intelligent in its own right, capable of altering organisms in response to their changing situations.

Jeremy Narby has calculated some amazing numbers in relation to this intelligent DNA. If the DNA packed into the invisibly small nuclei of each of our cells (along with protein and water) were stretched out, it would be about six inches long. End to end, the DNA of our several trillion cells would extend so far that it would take a jet plane traveling one thousand kilometers per hour over two centuries to reach its end! After this surprising result, Narby calculated that a single handful of living earth contains more DNA than that of our entire bodies—because bacteria are packed far more closely in soil than cellular nuclei are in our bodies.

Consider nature once more in this light—the entire surface of Earth covered and penetrated by intelligently self-organizing and reorganizing DNA in this almost inconceivable quantity. Is it truly the language of life through which cosmic consciousness expresses itself in 'material' worlds? Will we find it to be a common language throughout our co-created universe?

O   O   O

Not long after the mechanical theory of evolution was proposed by scientists, it was opposed by some philosophers, including Henri Bergson, Pierre Teilhard de Chardin, and Alfred North Whitehead, who worked out organic models or theories in which life is seen as inherent in the universe. Bergson opposed the idea of purpose in nature, but proposed the existence of a mysterious life force that is separate from and struggling with matter in an attempt to organize it. Scientists tended to reject his model because there was no room for a 'life force' in their worldview. Teilhard de Chardin, though he explained life as the natural evolution of self-organizing matter, saw evolution as purposive, leading—by way of mankind—to a "God-Omega point." Most scientists also rejected his work. Whitehead was ignored as too obscure for most scientists, but his talk of organics and God would have put them off in any case.

We have seen that at least a few scientists during the mechanical worldview era saw the Earth as alive or close to it. Darwin's younger Russian contemporary Vernadsky saw it as metabolically active; George Hutchinson of Yale University spread Vernadsky's ideas in America. James Lovelock and Lynn Margulis acknowledged geologist-physician James Hutton's concept of a living Earth as a forerunner to the Gaia hypothesis. Erich Jantsch's self-organizing universe called attention to the interplay of the largest and smallest events of cosmic matter/energy in producing the ever more complex systems of an essentially living universe. None of these scientific models included the notions of purpose or life force or God, as some of the philosophical ones did.

The mystery of life began clearing up as soon as we stopped thinking of the cosmos as a mechanical assembly of atomic, astral, and galactic parts—as soon as we began to see that cosmic 'parts' form themselves from—and are formed within—larger wholes, rather as species form themselves within, and are formed by, ecosystems in evolution. Is it not likely that life 'here below' behaves essentially the same way as things do 'above' in the greater cosmos, as the ancients claimed? If the universe is created by consciousness— cosmic consciousness differentiating into levels or layers of cosmic holarchy—and if an underlying sacred intelligent geometry is its means to create apparently physical worlds, then we should expect to find that it patterns itself in ways that repeat like fractals.

In Chapter 2 we saw that forms as simple as whirlpools— whether in water or in protogalactic clouds—take in matter, maintain their form in shaping its flow, and give off used matter to their environment—the essential pattern of any living thing's organization in relation to its surround. These vortex forms are as characteristic of subatomic particles as of galaxies and are found throughout the levels of nature familiar to us, for example in seashells, seedpods, rivers and tornadoes. Consistent with this basic model, all living holons gain their identity and relative autonomy by organizing their own form and function through a continual exchange—or re-creation—of matter while remaining dependent on their environment or larger holon to supply their resources and absorb their wastes. That is the pattern of the fractal cosmic and planetary dance.

Many scientists still unwilling to accept consciousness as the source of creation, rather than its late product, will continue for some time to vacillate between ideas of accident and purpose in attempts to explain how non-life becomes life. It took time to see matter and energy, wave and particle, time and space, as aspects of the same underlying unity; it may be even more difficult to see life and non-life as aspects of the cosmic process or as relative organizational states of cosmic matter. Scientists have no problem seeing the process of life in a puffball bursting to scatter its spores, but they do tend to have difficulty seeing the process of life in a star bursting to scatter *its* dust, for all that we are made of that dust.

The further development of a conscious, organic model of the universe clearly requires the cooperative efforts of biologists and physicists as well as other scientists, not to mention philosophers and spiritual leaders in a great co-creative process. With so much to learn and co-create as new stories, or worldviews, this is a most exciting and challenging time of discovery and new understanding. Most important in these developing efforts is the promise of seeing ever more clearly just how we humans fit into the great cosmic hol-archy of life and how we may learn to cooperate in creating its greatest health and fulfillment.

O   O   O

As we have seen, our historic worldviews—our images of who and what we are in relation to all nature—have been, on the one hand, limited by narrow perspective and, on the other, unbounded in egotism. For thousands of years we considered ourselves God's favorite creatures; then, when we had no more use for God, we saw ourselves as the pinnacle of natural evolution. In both views nature was ours to command and exploit as we liked. Only when we ourselves began suffering from the damage we had done to our environment did we begin to gain a more realistic view of ourselves as one species among still uncounted others on whom we depen In our cosmic worldview, sacred or secular, ours is a middle-sized planet circling an ordinary star at the edge of a common type of galaxy—a planet now in its middle age, about halfway through its expected life. Its oldest living creatures are bacteria, with an origin billions of years ago, but still the basis of its self-regulating systems and of all other life forms evolved since. Its newest living species are mammals that evolved only millions of years ago, though not a single species of early mammals survives today.

DNA is virtually the oldest thing in Earth's evolution still alive on its surface—propagating itself from the beginning in an unbroken chain, as surface rock transformed into endless creatures, who recycled it in turn into sediments that were subducted back into the magma of origin by great tectonic plates. All the while, DNA's species came and went, playing their roles and then disappearing, while it continued the dance.

There is little reason to think, from a biological perspective, that humans will change the pattern of species flux and survive for the rest of the life of our planet without further evolving. We haven't even the patience to wait for natural evolutionary changes, but are impatient to redesign our own DNA. What does this urgency signify? Do we sense that we are on the brink of a huge new leap in evolution? Do we think altering our own biology, rather than exercising our consciousness, is the way to get there?

No doubt we are a bold improvisation in Gaia's dance—a new kind of creature that may not survive very long at all if it doesn't learn to play a humbler and healthier role than it has thus far We produce and patent new plants and animals and create ever new kinds of artificial ecosystems.

Where are we headed in the arrogance of thinking we understand genetic engineering as well as those humble bacteria that invented, defined and refined it? In fact, we can only do it by enslaving *them* to carry out our intentions, as we have enslaved them to clean crude oil, to manufacture biodegradable plastics they invented long before us and to make them manufacture other particular substances they know how to make for us. But do we really want to eat plants we get them to insert poisons into every cell of to ward off insect predators, when nature long ago evolved plants healthy enough to keep them at bay in their own ways? Do we really want to clone ourselves to attain immortality, when our cloned sheep Dolly proved to age ten times faster than her peers?

Our rapid progress in biotechnology will likely get us into as much trouble as our nuclear-age mechanical technology if we don't make equal progress in understanding life systems and their dynamic ecological balance. Only if it is used with understanding of and respect for living systems can biotechnology offer the possibility of working *with* life *for* life.

O   O   O

On our planet, at least, we are indeed unique for the range of our conscious free choice of behavior, the range of our technological prowess, and our ability to foresee, plan, and act for the future. But the very evolution of such abilities on our own planet suggests they may well have evolved on many other planets around the universe. The same kind of matter exists everywhere and seems to undergo the same processes all over our cosmos. There is no good reason not to assume that planets come alive and evolve in essentially the same way everywhere. This assumption is fairly widespread among scientists now and underlies our ongoing efforts to reach and communicate with intelligent life in other star systems.

In most of these communications efforts, such as the radio telescope project CETI, scientists assume that intelligent species of other planets will have invented a mechanical technology composed of mathematics, machinery and electronics essentially similar to our own, and that they will not have progressed beyond it. This assumption is made in the belief that intelligence—the gathering and using of information to gain understanding—can grow to human proportions only through our kind of technology, through a similar co-evolution of conscious thought with the productive use of manipulative organs such as our hands. This idea rules out the possibility that creatures such as cetaceans could be as intelligent as humans, and it would be wise to feel less sure of ourselves on this score until we know more about the whales and dolphins.

It is interesting to consider that several of our own technological inventions—including sonar, diving equipment, insulated clothing, and high-speed long-range communications systems—all evolved within cetacean bodies, which function comfortably at widely varying pressures and temperatures as they move freely about the seas covering most of the Earth's surface. We have no present way of knowing how far such naturally evolved internal biotechnology can go or how it can be used by highly evolved brains—or whether it could contribute to the formation or function of living systems beyond our planetary Gaia.

There is a whole literature about and by people who have had demonstrable telepathic communications with cetaceans.

Indigenous humans use telepathy routinely to communicate complex thoughts and concepts with each other, as well as with other species. One such tribe's use of telepathy is described in detail by Petru Popescu, in his story about National Geographic photographer Loren McIntyre's experiences with the "cat people" of the Amazon, who had been thought extinct until he stumbled on them while lost. It is of special interest that these so-called primitive people have sophisticated understanding of both practical linear time and a deeper non-linear understanding of radial time as physicists only now begin to comprehend it.

Harvard psychiatrist John Mack, whose credibility was put to severe academic test and passed, has interviewed patients and others around the world about their experiences with extra-terrestrials. His findings show almost universally that ET communication is telepathic and so prolific in 'downloading' that the people involved can only grasp and communicate small parts of it.

Many people with ET contacts believe that galactic and cosmic life systems larger than those confined to single planets already exist, and our very popular science fiction films have reflected that belief for some time. A galactic living system would require communications and space travel, but most likely not of the sort we have imagined—radio telescopes and great space ships supplied for generations of physical life forms.

If live planets are the cells of such galactic systems, their living creatures may not have to come into physical contact with one another, any more than the mitochondria of our liver cells need to meet the mitochondria of our bone cells. A common information-communications system may unite them effectively.

Clearly linear time and physical space as we have understood them, preclude any kind of efficient or reasonable space travel.

Even the exchange of messages coming one at a time across light years of space now look like a crude and primitive idea. Does that mean that space travel and communications are impossible, or simply that our concepts of time and space were wrong? Non-locality, radial time, creation by consciousness and other concepts new to western science are rapidly changing our assumptions on how our universe works, and opening doors to our long-dreamed of contacts with other beings of our cosmos. In fact, once we see that these things are possible, as Joe Firmage points out, we will be less reluctant to believe they are happening, and perhaps always have been throughout human history.

O   O   O

Before we consider how we might become part of an interplanetary or intergalactic life system, shouldn't we consider how we might look to other members of such a system? We have seen the Gaian system and ourselves from various perspectives within or just outside it. Let us now try looking at ourselves from the point of view of some intelligent species from another star system that can observe us Learning what we are and what we are up to, would they consider us an intelligent form of life?

Surely it would strike them as most peculiar that we destroy the environment on which we depend. No intelligent species would knowingly pollute its air, water, and soil to the point of endangering itself. It would hardly cram itself into communities of concrete that sealed the species off from natural processes and made its air unbreathable with its own wastes when there was plenty of space on the planet and ways to avoid creating the pollution.

They would surely wonder why we destroy the natural ecosystems of our planet to grow our own food without preserving variety and recycling water and nutrients to prevent the land from turning into deserts. They would note that we deliberately overload our planet's delicately balanced atmosphere and thus overheat the planet itself with carbon dioxide as we hack down our tropical forests and destroy the coastal areas that might reestablish the balance. They would wonder how we could know that our whole planet's life system is driven effectively by solar energy and not use this safe energy source extensively ourselves, if we cannot yet tap deeper sources such as zero-point energy.

Could they consider us intelligent after they see that we are quite capable of providing a comfortable life for all humans, yet choose to devote enormous resources to escalating an arms race rather than using them for the well-being of our people? Would an intelligent species overproduce food for some of its members while millions of its young die annually of starvation? Would our observers not wonder that the males of our species had long ago declared the female half of the species inferior, largely excluding them from their former positions of authority and management, at the same time often shifting social priorities from life-giving to life-taking?

They would see us systematically exterminating other species and whole ecosystems, even the other large-brained creatures most like ourselves, such as dolphins and whales, elephants, chimpanzees, and gorillas, though they are peaceful and inoffensive to humans. They would see our leaders, many of whom hold their positions by popular consent, maintaining hostilities that threaten nuclear warfare, which would destroy all parties and create a planetary nuclear winter. What would they think of our popular Star Wars weapons cult and of our use of that name for a weapons system?

If extraterrestrial species developed technologies essentially similar to ours, they would sooner or later discover nuclear energy or even more sophisticated means of destruction. If they were warring species at the time of such discoveries, they would have faced crises similar to our own—crises of choice between species suicide or species cooperation. One might think any species that has survived such discoveries would have learned peaceful cooperation or blasted itself out of existence.

This leaves our expectations of alien belligerence less likely, and perhaps even embarrassing, if it comes to a meeting. Are we a cosmic anomaly—a species once mutually consistent with its ecosystem that became rather suddenly, over a period of a few thousand years, hostile and destructive to itself and its living planet? May we attribute this peculiarity to the temporary condition of youth—to unprecedented freedom during the heady stage of species adolescence? Have our observers reason to hope we shall soon grow up and become wiser?

In the stories of ET encounters Dr. Mack hears from his patients, the ET's are indeed appalled at the destructiveness of human behavior, and their most common plea to humans is to take care of their planet while they still can. What is the evidence they might see in our favor—that might give them hope we *can* behave as an intelligent species?

We have begun to understand and be concerned with ecological balance; we are beginning in at least some areas to protect endangered species, reduce our pollution, and give the rest of nature a chance to clean up and restore its balance. Unfortunately these constructive efforts are very far from balancing our continuing destruction. Still, there is hope that we may increase these efforts dramatically when we finally recognize ourselves as a living system—part of a living being whose delicate balance is tailored to our needs, but which may eliminate us if we force it to reorganize itself to cope with our reckless destruction of that balance.

The threat of all-out nuclear war actually lessened dramatically as we took to heart our scientists' predictions of its dire consequences for all humanity, but genuine disarmament has not happened among the great powers, and renegade nuclear weapons in smaller nations proliferate, ever more difficult to track. This suggests that we do respond to our knowledge by changing our ways, but we need to speed this process.

Our technology is well on the way out of the industrial age of heavy steel and polluting fossil fuels, into the information age of lightweight transistor technology, the Web, and benign energy sources such as sunlight, wind, waves, hydrogen and alcohol, with water fuel cells and zero-point energy on the immediate horizon. If we put our future welfare ahead of immediate profit motives, turning corporations into living systems that make only consumable and recyclable products, this transition could be completed rapidly and effectively, as we said in the last chapter.

Agricultural research, as we also saw, is beginning to look more seriously at older, even ancient, methods of natural pest control and crop rotation and variation, recognizing that monocultures sustained by chemicals are rapidly destroying the land and polluting the waterways. The shift to organic agriculture is already underway.

Our worldwide economic system, our transportation and communications technology, our information revolution, have bound us into a body of humanity that is now being pushed for the sake of its survival to evolve from competition to cooperation among nations and with our environment. We see that the integrative evolutionary forces, which produced protists and multicelled creatures, are now pushing us to complete our own organizational task. We see that we only prolong and aggravate our biggest problems by resisting this evolution with habitual fears and hostile competition. We can use our gift of freedom to make up for our lack of innate limits to territorial and aggressive behavior by channeling these into constructive negotiation and sharing, as we are practicing on the Web.

We can recognize that the strength and resilience of living systems lies in their diversity, and stop trying to make ourselves all alike. We can analyze and reorganize efforts such as the United Nations, which are as critical to the organization of the body of humanity as were nuclei to eukaryotes and brains to animals.

We can end the sexual inequality that is not merely an injustice to women but a deplorable waste of half our species' talents for creative management and nurture of the species as a whole. As long as women remain a tiny minority in positions of human leadership, they will be pressed on the whole to conform to the established male model of society based on bureaucracy, top-down command and control organizations, competition, conquest, and profit. Only when women assume equal leadership will they be free to express effectively their abilities in organic organization, networking, cooperation, nurture, and mutual benefit.

Certainly we would be foolish to continue our environmental destruction and our hostilities to the point of possible extinction when we are surrounded by clues to exciting, creative, natural solutions for our greatest problems. If we accept ourselves as an adolescent species in crisis and face the challenge of taking on mature responsibility for our freedom with courage and enthusiasm, we stand an excellent chance of growing up and reaping the benefits of maturity.

O   O   O

How rewarding it would be if, when we openly communicate with extraterrestrials we could do so having solved our great problems of inequality, hunger, pollution, devastation of ecosystems and nuclear threat. We would then be in a wonderful position to face new challenges with appropriate pride in our species. Perhaps we will be called on to protect Gaia by detecting and diverting massive meteors or planetoids; perhaps if we come to understand and help foster Gaia's healthy ecosystems, we will be able to bring life to Mars or some Moons of our solar system, thus spreading Gaia's seed Whatever our dreams of such roles, or of cooperating in a larger cosmic life system, our immediate task is still here at home. On the whole, there seems to be good reason to believe our recklessly egotistical and destructive phase is coming to an end with new knowledge that leads us back to ancient wisdom. We are capable of regaining our reverence for life, of replacing the drive to conquer with the will to cooperate, of remaking our engineered institutions, including our corporations, into living systems.

The more we learn about nature, including human nature, the more we can see that our living parent planet and our whole living cosmos are far more beautiful and awesome in the reality of their self-creation than is any myth we made as we struggled to develop our knowledge. At last our scientific and religious quests can merge in the recognition that conscious, sacred, self-creating nature, both Gaian and cosmic, is our physical and spiritual source, the wellspring of our ancient inspiration to love, and the experienced guide we have always sought—the guide we need more than ever now that we stand on the brink of maturity.

Bibliography

Abraham, Ralph H. 1981. "Dynamics and Self-Organization." In F.F.Yates, *Self-Organizing Systems: The Emergence of Order*. New York: Plenum.

Andruss, Van; Plant, Christopher; Plant, Judith; Wright, Eleanor, Eds. 1990. *Home: A Bioregional Reader*. Philadelphia: New Society Publishers.

Augros, Robert M., and George N. Stanciu. 1986. *The New Story of Science*. New York: Bantam.

Balandin, R. K. 1982. *Vladimir Vernadsky*. Outstanding Soviet Scientists Series. Moscow: Mir Publishers.

Barnett, Lincoln. 1957. *The Universe and Dr. Einstein*. New York: Bantam.

Barrow, J. D., and F. J. Tipler. 1986. The *Anthropic Cosmological Principle*. Oxford, England: Clarendon Press.

Bateson, Gregory. 1972. *Steps to an Ecology of Mind*. New York: Ballantine.

Bateson, Gregory. 1979. *Mind and Nature: A Necessary Unity*. New York: Dutton.

Benyus, Janine. 1997 *Biomimicry: Innovation Inspired by Nature*. NY: William Morrow and Company.

Bergson, Henri. 1911; reprint, 1983. *Creative Evolution*. Lanham, Md: University Press of America.

Berry, Thomas. 1988. *The Dream of the Earth*. SanFrancisco: Sierra Club Books.

Berry, Thomas. 1987. "Thomas Berry: A Special Section."*Crosscurrents* 38 (2, 3).

Bohm, David. 1957. *Causality and Chance in Modern Physics.* New York: Harper Torchbooks.

Bohm, David. 1980. *Wholeness and the Implicate Order.* London: Routledge & Kegan Paul.

Boon, Jaap. 1984. "Tracing the Origin of Chemical Fossils in Microbial Mats." In Y. Cohen, R. W. Castenholz, and H. O. Halvorson. *Microbial Mats: Stomatolites.* New York: Alan R. Liss.

Boorstin, Daniel J. 1985. *The Discoverers: A History of Man's Search to Know His World and Himself.* New York: Vintage.

Briggs, John P., and F. David Peat. 1986. *Looking Glass Universe.* New York: Simon & Schuster.

Bronowski, J. 1974. *The Ascent of Man.* Boston: Little, Brown & Co.

Brown, L. R., E. C. Wolf, S. Postel, W. U. Chandler, C. Flavin, and C. Pollock. 1985. "State of the Earth," *Natural History*, April.

Cajete, Gregory. 1994. *Look to the Mountain: An Ecology of Indigenous Education.* Kivaki Press: Durango, CO.

Calder, Nigel. 1978. *The Key to the Universe: A Report on the New Physics.* New York: Penguin.

Campbell, Joseph. 1968. *The Masks of God: Creative Mythology.* New York: Penguin.

Cannon, Walter. 1939. *The Wisdom of the Body.* New York: Norton.

Capra, Fritjof. 1975. *The Tao of Physics.* Berkeley: Shambala.

Capra, Fritjof. 1982. *The Turning Point: Science, Society and the Rising Culture.* New York: Simon & Schuster.

Charlson, R., J. Lovelock, M. Andreas, and S. Warren. 1987. "Oceanic Phytoplankton, Atmospheric Sulfur, Cloud Albedo and Climate." *Nature* 326:655.

Chorover, Stephen. 1976. *From Genesis to Genocide*. Cambridge, MA: M.I.T. Press.

Daly, Herman E., and Cobb, John B. Jr. 1989. *For the Common Good: Redirecting the Economy Toward Community, the Environment, and a Sustainable Future*. Boston: Beacon Press.

Darwin, Charles, and T. H. Huxley. 1983. *Darwin and Huxley: Autobiographies*. Edited by Gavin de Beer. Oxford: Oxford University Press.

Dawkins, R. 1976. *The Selfish Gene*. Oxford: Oxford University Press.

Dostoevski, Feodor. 1879; reprint, 1977. *The Brothers Karamazov*. New York: Penguin.

Dunhamn, Barrows. 1960. *Thinkers and Treasurers*. New York: Monthly Review Press.

Durant, Will. 1961. *The Story of Philosophy*. New York: Simon & Schuster.

Easlea, Brian. 1983. *Fathering the Unthinkable: Masculinity, Scientists and the Nuclear Arms Race*. London: Pluto Press.

Elgin, Duane. 1981, 1993. *Voluntary Simplicity: An Ecological Lifestyle that Promotes Personal and Social Renewal*. NY: Bantam.

Elgin, Duane. 1993. *Awakening Earth: Exploring the Evolution of Human Culture and Consciousness*. NY: William Morrow and Co.

Ehrlich, Paul and Ann. 1975. *The Population Bomb*. Mattituck, NY:Amereon, River City Press.

Einstein, Albert. 1950. *Out of My Later Years*. Scranton, Pa: Philosophical Library.

Einstein, Albert. 1961. *Relativity: The Special and General Theory*. New York: Crown.

Eisler, Riane. 1987. *The Chalice and the Blade*. San Francisco: Harper & Row.

Ereira, Alan. 1991. *Message from the Heart of the World: the Elder Brother's Warning*. London: BBC Film Studios.

Firmage, Joseph

Fischer, Dietrich. *Non-Military Aspects of Security: A Systems Approach Report to the United Nations Institute for Disarmament Research (UNIDIR)*.

Fleischaker, G. R. 1988. *Autopoiesis: System Logic and Origins of Life*. Ph.D. dissertation, Boston University.

Fox, Michael. 1992. *Superpigs and Wondercorn: The Brave New Technology of Biotechnology and Where It All May Lead*. New York: Lyons & Burford.

Gates, Jeff. 1998. *The Ownership Solution: Toward a Shared Capitalism for the 21st Century*. Reading, MA: Addison-Wesley

Gates, Jeff. 2000. *Democracy at Risk*. Perseus Books.

Gehrz, R. D., D. C. Black, and P. M. Solomon. 1984. "The Formation of Stellar Systems from Interstellar Molecular Clouds," *Science*, 25 May.

Gimbutas, Marija 1975. "Figurines of Old Europe (6500-3500 B.C.)." *Transactions of the International Valcamonica Symposium on Prehistoric Religions*. Capo di Ponte, Brescia, Italy: Edizione del Centro.

Gimbutas, Marija 1980. "The Temples of Old Europe," *Archaeology*, November-December.

Gimbutas, Marija 1981. "The Monstrous Venus of Prehistory, or Goddess Creatrix." *Comparative Civilizations Review*, 7 (Fall 1981).

Gimbutas, Marija 1982. "Megalithic Tombs of Western Europe and Their Religious Implications." *Quarterly Review of Archaeology* 6(3).

Gimbutas, Marija 1982. *The Goddesses and Gods of Old Europe, 7000-3500 B.C.* Berkeley: University of California Press.

Goerner,Sally, 1999. *After the Clockwork Universe: The Emerging Science and Culture of Integral Society.* Edinburgh: Floris Books.

Goldsmith, Edward. 1993. *The Way: An Ecological World-view.* Boston: Shambhala.

Goodall, Jane. 1983. *In the Shadow of Man.* Boston: Houghton Mifflin.

Gore, Al. 1992. *Earth in the Balance: Ecology and the Human Spirit.* NY: Houghton Mifflin.

Gould, S. J. 1977. *Ever Since Darwin: Reflections in Natural History.* New York: Penguin.

Gould, S. J. 1984. *The Mismeasure of Man.* New York: Penguin.

Graves, Robert. 1957. *The Greek Myths.* New York: Penguin.

Gray, Michael W. 1983. "The Bacterial Ancestry of Plastids and Mitochondria." *BioScience* 33(11).

Grene, David, and Richard Lattimore, eds. 1956. *The Complete Greek Tragedies.* Chicago: University of Chicago Press.

Grey, Victor. 1997. *Web Without a Weaver.* Concord, CA: Open Heart Press.

Haldane, J.B.S. 1985. *On Being the Right Size and Other Essays.* Edited by J. M. Smith. Oxford: Oxford University Press.

Hawken, Stephen 1988. *A Brief History of Time: From the Big Bang to Black Holes.* NY: Bantam.

Heinberg, Richard. 1999. *Cloning the Buddha: The Moral Impact of Biotechnology.* Wheaton, IL: The Theosophical Publishing House.

Heisenberg, Walter. 1975. *Across the Frontiers*. New York: Harper Torchbooks.

Henderson, Hazel. 1991. *Paradigms in Progress: Life Beyond Economics*. Indianapolis: Knowledge Systems, Inc.

Ho, M. W., and P. T. Saunders, eds. 1984. *Beyond Neo-Darwinism: An Introduction to the New Evolutionary Paradigm*. London: Academic Press.

Ho, M. W., P. T. Saunders, and S. W. Fox. 1986. "A New Paradigm for Evolution," *New Scientist*, 27 Feb.

Hodge, Helena Norberg. 1991. *Ancient Futures: Learning from Ladakh*. San Francisco: Sierra Club Books. Video available from ISEC, P.O. Box 9475, Berkeley, CA 94709.

Hooker, Michael, ed. 1978. *Descartes*. Baltimore: Johns Hopkins University Press.

Huston, Perdita. 1982. *Message from the Village*. New York: The Epoch B Foundation.

Ingber, Donald E.. 1998. "The Architecture of Life," *Scientific American*. January.

Jaikaran, Jacques. 1992. *Debt Virus: A Compelling Solution to the World's Debt Problems*. Lakewood, Colorado: Glenbridge Publishing, Ltd:

Jantsch, E. 1980. *The Self-Organizing Universe: Scientific and Human Implications of the Emerging Paradigm of Evolution*. Oxford, England: Pergamon Press.

Jantsch, E., and C. H. Waddington. 1976. *Evolution and Consciousness: Human Systems in Transition*. Reading, Mass: Addison-Wesley.

Kaiser, Rudolf. 1991. *The Voice of the Great Spirit: Prophecies of the Hopi Indians*. Boston: Shambala.

Kasting, J., O. Toon, and J. Pollack. 1988. "How Climate Evolved on Terrestrial Planets." *Scientific American*, February, p. 90.

Kerr, R. A. 1988. "No Longer Willful, Gaia Becomes Respectable," *Science*, April Kirk, G. S., J. E. Raven, and M. Schofield. 1983. *The Pre-Socratic Philosophers.* Cambridge, England: Cambridge University Press.

Kitto, H.D.F. 1978. *Greek Tragedy: A Literary Study.* London: Methuen.

Kitto, H.D.F. 1979. *The Greeks.* New York: Penguin.

Koestler, A. 1978. *Janus: A Summing Up.* London: Pan Books.

Kolluru, Rao V. *1994 Environmental Strategies Handbook: A Guide to Effective Policies & Practices.* McGraw-Hill: NY

Korten, David. 1999. *The Post-Corporate World: Life After Capitalism.* San Francisco: Berrett-Koehler Publishers and Kumarian Press.

Lamb, Bruce. 1985. *Rio Tigre and Beyond: The Amazon Jungle Medicine of Manuel Cordova.* Berkeley: North Atlantic Books.

Lapo, A. V. 1982. *Traces of Bygone Biospheres.* Moscow: Mir Publishers.

Lappe, Francis Moore, and Joseph Collins. 1977. *World Hunger: Ten Myths.* San Francisco: Free Spirit Press, Institute for Food and Development Policy.

Lazcano, A. 1986. "Prebiotic Evolution and the Origin of Cells." In *Origin of Life and Evolution of Cells.* Edited by L. Margulis, R. Guerrero, and A. Lazcano. *Treballs de la Societat Catalana de Biologia*, vol. 39.

Lazlo, Ervin. 1978. *The Inner Limits of Mankind.* Oxford: Pergamon Press.

Lazlo, Ervin. 1994. *The Choice: Evolution or Extinction.* NY: Tarcher/Putnam.

Lemkow, Anna F. 1990. *The Wholeness Principle*. Wheaton IL: The Theosophical Publishing House.

Levine, Rick, Christopher Locke, Doc Searls & David Weinberger. 1999. *The Cluetrain Manifesto: The End of Business as Usual*. Cambridge: Perseus Books.

Light, Luise. 1994. "*Eating Dangerously: The Risk of Food Poisoning from Bacteria in Meat and Poultry and What We Can Do About It.*" Washington D.C: Institute for Science in Society.

Lilly, John C. 1975. *Lilly on Dolphins*. Garden City, NY: Doubleday/ Anchor.

Lovelock, J. E. 1972. "Gaia As Seen through the Atmosphere." *Atmospheric Environment* 6 (579).

Lovelock, J. E. 1982. *Gaia: A New Look at Life on Earth*. Oxford: Oxford University Press.

Lovelock, J. E. 1986. "Geophysiology: The Science of Gaia." *The New Scientist*. London, n.d.

Lovelock, J. E. 1988. *The Ages of Gaia: A Biography of Our Living Earth*. New York: Norton.

Lovelock, J.E. 1991 *Healing Gaia: Practical Medicine for the Planet*. NY: Norton.

Lovelock, J. E., and L. Margulis. 1984. "Gaia and Geognosy." In M. B. Rambler. *Global Ecology: Towards a Science of the Biosphere*. London: Jones and Bartlett.

Lyons, Oren. et al. 1992. *Exiled in the Land of the Free*. Democracy, Indian Nations and the United States Constitution. Santa Fe, NM: Clear Light.

Mack, John. 1994. *Abduction*. NY: Ballantine.

Mander, Jerry. *In the Absence of the Sacred: The Failure of Technology and the Survival of the Indian Nations*. 1991. San Francisco: Sierra Club Books.

Margulis, L. 1981. *Symbiosis in Cell Evolution*. San Francisco: Freeman.

Margulis, L. 1982. *Early Life*. Boston: Science Books International.

Margulis, L., and R. Guerrero. 1986. Not <origins of life> but <evolution in microbes>." In *Origin of Life and Evolution of Cells, Treballs de la Societat Catalana de Biologia*, vol. 39. Edited by L. Margulis, R. Guerrero, and A. Lazcano.

Margulis, L., and D. Sagan. 1986. *Origins of Sex: Three Billion Years of Genetic Recombination*. New Haven, Conn: Yale University Press.

Margulis, L., and D. Sagan. 1987. *Microcosmos: Four Billion Years of Evolution from Our Microbial Ancestors*. London: Allen & Unwin.

Margulis, L. and D. Sagan.1987 *What is Sex?* New York: Simon & Schuster Maturana, Humberto R., and Francisco J. Varela 1987. *The Tree of Knowledge: The Biological Roots of Human Understanding*. Boston: Shambala

McKibben, Bill. 1995 *Hope, Human and Wild:True Stories of Living Lightly on the Earth*. New York: Little, Brown & Company.

McLaughlin, Corrinne and Davidson, Gordon. 1994 *Spiritual Politics*. NY: Ballantine.

Mellaart, James. 1975. *The Neolithic of the Near East*. New York: Scribner's.

Mellaart, James. 1967. *Catal Huyuk*. New York: McGraw-Hill.

Merchant, Carolyn. 1980. *The Death of Nature: Women, Ecology and the Scientific Revolution*. San Francisco: Harper & Row.

Mollison, William. 1992 *Permaculture: A Practical Guide for a Sustainable Future*. Cedar Crest, New Mexico: Tagari Publications of Permaculture Services International.

Narby,Jeremy. 1998. *The Cosmic Serpent: DNA and the Origins of Knowledge.* NY: Tarcher/Putnam.

Newman, James R., ed. 1956. *The World of Mathematics*, 4 vols. New York: Simon & Schuster.

Odum, Eugene P. 1983. *Basic Ecology.* Philadelphia: Saunders College Publishing.

Ong, W. 1982. *Orality and Literacy: The Technologizing of the Word.* London: Methuen.

Pankow, Walter. 1979. "Openness as Self-Transcendence." In E. Jantsch and C. H. Waddington, eds. *Evolution and Consciousness.* Reading, MA: Addison-Wesley.

Pearce, F. 1988. "Gaia: A Revolution Comes of Age," *New Scientist*, 17 March.

Petzinger, Thomas. 1999. *The New Pioneers: The Men and Women Who are Transforming the Workplace and the Marketplace.* NY: Simon & Schuster.

Plant, Christopher and Plant, Judith, Eds. 1991. *Green Business: Hope or Hoax.* Philadelphia: New Society Publishers.

Plant, Christopher and Plant, Judith, Eds. 1992. *Putting Power in its Place.* Philadelphia: New Society Publishers.

Plato. Reprint, 1956. *Great Dialogues of Plato.* Translated by Philip G. Rouse. New York: New American Library.

Popescu, Petru. 1991 *Amazon Beaming.* NY: Vilking.

Posey, Darrell Posey. 1992. "Indigenous Knowledge in the Conservation and Utilization of World Forests." Discrimination Against Indigenous Peoples. UN Economic and Social Council Distr. General E/CN.4/Sub.2/1992/31/Add.1

Postgate, J. 1988. "Gaia Gets Too Big for Her Boots," New Scientist, 7 April

Prigogine, Ilya, and Isabelle Stengers. 1984. *Order Out of Chaos: Man's New Dialogue with Nature*. New York: Bantam.

Puthoff, Harold. 1990. "Everything for Nothing." New Scientist. 28 July.

Rifkin, Jeremy. 1980. *Entropy: A New World View*. The Viking Press: New York

Roberts, Jane. 1981, 1995. *The Individual and the Nature of Mass Events*. Amber-Allen, SanRafael, CA.

Roberts, Jane. 1998, 1999. *The Early Sessions*, vols 1-6, NY: New Awareness Network, Inc. (P.O.Box 192, Manhassett, NY 11030).

Robin, Vicki and Dominguez, Joe. 1993. *Your Money or Your Life: Transforming Your Relationship With Money and Achieving Financial Independence*. NY: Penguin

Rothstein, Jerome. 1985. "On the Scientific Validity and Utility of the Living Earth Concept and Its Further Generalization." Symposium, *"Is the Earth a Living Organism?"* Amherst, Mass: University of Massachusetts.

Russell, Bertrand. 1961. *History of Western Philosophy*. London: Allen & Unwin.

Russell, D. A. 1979. "The Enigma of the Extinction of the Dinosaurs." *Annual Review of Earth Planetary Science 7*: 163-82.

Russell, Peter. 1992, 1998. *Waking Up in Time*. Novato, CA: Origin Press.

Sagan, D., and Margulis, L. 1983. The Gaian Perspective of Ecology. *The Ecologist* 13(5).

Sahlins, Marshall. 1972. *Stone Age Economics*. NY: Aldine de Gruyter.

Sahtouris, Elisabet. 1997. "The Biology of Globalization," in *Perspectives on Business and Global Change , Journal of the World Business Academy*. September.

Sahtouris, Elisabet and Harman, Willis. 1998. *Biology Revisioned*. Berkeley: North Atlantic Publishers.

Sahtouris, Elisabet, with Liebes, Sid and Swimme, Brian. 1998. *A Walk Through Time: From Stardust to Us*. NY: Wiley.

Schneider, S. H. & Boston, P. 1993. *Scientists on Gaia*. Cambridge, MA: The MIT Press.

Scientific American. 1983. "The Dynamic Earth." September.

Shiva, Vandana. 1989. *The Violence of the Green Revolution*. Dehra Dun, India.

Shiva, Vandana. 1997. *Biopiracy: The Plunder of Nature and Knowledge*. Boston: South End Press.

Skinner, B. F. 1975. *Beyond Freedom and Dignity*. New York: Bantam.

Sonea, S., and M. Panisset. 1983. *A New Bacteriology*. Boston: Jones & Bartlett.

Spretnak, Charlene. 1981. *Lost Goddesses of Ancient Greece*. Boston: Beacon Press.

Stone, Merlin. 1976. *When God Was a Woman*. New York: Harcourt Brace Jovanovich.

Swimme, Brian. 1984. *The Universe Is a Green Dragon*. Santa Fe: Bear.

Teilhard de Chardin, Pierre. Reprint, 1959. *The Phenomenon of Man*. London: William Collins.

Thomas, Lewis. 1975. *The Lives of a Cell: Notes of a Biology Watcher*. New York: Bantam.

Toffler, Alvin. 1980. *The Third Wave*. London: William Collins.

Toynbee, Arnold. 1972. *A Study of History*. Oxford: Oxford University Press.

Varela, F. J., H. R. Maturana, and R. Uribe. 1974. "Autopoiesis: The Organization of Living Systems, Its Characterization and a Model." *Biosystems 5:* 187-96.

Volk, Tyler. 1998. *Gaia's Body: Toward a Physiology of Earth.* NY: Springer-Verlag NY Inc.

Vrooman, J. R. 1970. *Rene Descartes.* New York: Putnam.

Waldrop, M. Mitchell. 1992. *Complexity: The Emerging Science of Order and Chaos.* New York: Simon & Schuster.

Wallace, R. A., J. L. King, and G. P. Sanders. 1984. *Biosphere: The Realm of Life.* London: Scott Foresman.

Watson, Lyall 1979. *Lifetide: A Biology of the Unconscious.* London: Hodder & Stoughton.

Watts, Alan 1966. *The Book: On the Taboo Against Knowing Who You Are.* NY: Collier Books.

Weatherford, Jack. 1988. *Indian Givers.* NY: Crown Publishers.

Whitehead, A. N. 1926. *Science and the Modern World.* New York: Macmillan.

Whitehead, A. N. 1979. *Process and Reality.* New York: Free Press.

Whorf, Benjamin Lee. 1956. *Language, Thought and Reality: Selected Writings.* Edited by John B. Carroll. Cambridge, Mass.: M.I.T. Press.

Wilber, Ken. 1998 *The Marriage of Sense and Soul: Integrating Science and Religion.* NY: Random House.

Wilson, E. O. 1975. *Sociobiology.* Cambridge, Mass.: Harvard University Press.

Wolf, Fred Alan. 1993. *The Eagle's Quest.* NY: Simon & Schuster Pocketbooks.

World Game Institute. *Doing the Right Things*, available from the World Game Institute, 3508 Market Street, Philadelphia, PA 19104; Tel: 215/387-0220.

Xenophon. Reprint, 1979. *Recollections of Socrates and Socrates' Defense before the Jury*. Translated by Anna J. Benjamin. Indianapolis: Bobbs-Merrill.

About the Author

Elisabet Sahtouris, Ph.D. is an evolution biologist, futurist and author/lecturer who has lived in the USA, Greece and Peru. She has taught at MIT, the University of Massachusetts and CIIS. She contributed to the NOVA/HORIZON series at WGBH-TV Boston and has been a UN consultant on indigenous peoples. Her work focuses on bringing principles of living systems into the corporate world, governments and other organizations, and into the process of globalization. She has two children and three grandchildren. Her recent books are *Biology Revisioned*, a dialogue with Willis Harman, North Atlantic Publishers 1998, and *A Walk Through Time: From Stardust to Us*, Wiley 1998.

She is available for consulting and speaking engagements.
E-mail: elisabet@sahtouris.com
Websites: www.sahtouris.com; www.ratical.org/lifeweb